BIOMEDICAL ETHICS REVIEWS

Edited by *James M. Humber and Robert F. Almeder*

BIOMEDICAL ETHICS REVIEWS

Edited by *James M. Humber and Robert F. Almeder*

BIOMEDICAL
E T H I C S
R E V I E W S

STEM CELL RESEARCH

Edited by

James M. Humber

and

Robert F. Almeder

Georgia State University, Atlanta, Georgia

Humana Press • Totowa, New Jersey

The Library of Congress has cataloged this serial title as follows:

Stem cell research / edited by James M. Humber and Robert F. Almeder.
     p. ; cm. -- (Biomedical ethics reviews ; 2004)
Includes index.
    ISBN 1-58829-401-3 (alk. paper)
    1. Stem cells--Research--Moral and ethical aspects.
    [DNLM: 1. Embryo Research--ethics. 2. Stem Cells.   QS 620 S824
2004]   I. Humber, James M. II. Almeder, Robert F. III. Series.
    QH587.S7226 2004
      174'.28--dc21
                                  2003009570

Preface

Today, few doubt that human embryonic stem cell research possesses the potential for significantly reducing the suffering of those inflicted with such severely debilitating diseases as Parkinson's disease. It is this potential that has led many to argue that stem cell research should proceed with all deliberate speed and with as few encumbrances as possible. Others demur. These individuals acknowledge that human embryonic stem cell research might lead to beneficial results, but nevertheless insist that research of this sort is of such a nature that it must be considered morally suspect. More specifically, they argue that the morally troublesome character of human embryonic stem cell research provides sufficient grounds for believing that this type of experimentation should be proscribed or, at the very least, severely restricted and regulated. At present, the debate between the pro- and anti-stem cell research factions continues to be waged without showing any signs of imminent resolution. The purpose of *Stem Cell Research* is to introduce readers to the principal arguments used by both sides in the dispute.

When examining the arguments that have been advanced for and against the moral propriety of stem cell research, certain issues have come to be viewed as preeminent. All of the chapters in this text focus on issues of this sort. Among these issues are the following: What reason (or reasons) are there for believing that the materials used in human embryonic stem cell research should (or should not) be included as full members of the moral community and thus be accorded all of the protections ordinarily extended to such individuals? Does the present debate over stem cell research overemphasize the importance of determining the moral status of the human embryo? Would it be morally wrong to create embryos with the human genome, specifically to secure embryonic stem cells for use in research, knowing that these embryos will be destroyed in the process? Would it be wrong to experiment with stem cells taken from human embryos that were not originally created to produce materials for stem cell research, knowing that the experimentation will lead to the destruction of the embryos? Would it be wrong to experiment with cells taken from human embryos that were not created to produce materials for stem cell research when we know those embryos are about to be destroyed and discarded? Does the current use of human embryonic stem cells in research devalue humanity by treating these materials as mere "commodities?" Finally, are the current regulations on stem cell research unjust because they alienate women from their reproductive labor?

Stem Cell Research is the 21st annual volume of *Biomedical Ethics Reviews*, a series of texts designed to review and update the literature on issues of central importance in bioethics today. It is also the last volume in the series that the current editors will prepare for publication. James Humber is retiring, and Robert Almeder is engaged in so many academic pursuits that he feels he lacks the time to continue as sole editor for the series and do full justice to the job. Still, readers should know that Humana Press is committed to the continued publication of volumes in the *Biomedical Ethics Reviews* series and is currently engaged in the process of selecting a new editor for the series.

We, the editors, have enjoyed working with all of the employees of Humana Press with whom we have had contact over the years. We would especially like to thank Thomas L. Lanigan, president of Humana Press, for his unwavering support, trustworthiness, and invaluable advice. The idea for *Biomedical Ethics Reviews* was his, and we deeply appreciate the fact that he selected us to serve as the original editors for the series. We hope his expectations for *Biomedical Ethics Reviews* have been fulfilled, and are confident that whomever he selects to edit future volumes of *Biomedical Ethics Reviews* will make us proud to have our names forever associated with the series.

In closing, one final comment seems in order. Never, in any of the pages of *Biomedical Ethics Reviews* that we have edited in the preceding 21 years, have we taken note of any of the contributions and sacrifices our wives have made to provide us with the time necessary to do our work. We know they expect no thanks for their efforts, and suspect that they know us well enough to realize how much we appreciate their support despite the fact that we have been silent for 21 years. Clearly, some formal statement of thanks is called for, and in recompense for our past errors of omission we would like to dedicate this volume of *Biomedical Ethics Reviews* to our wives, Helene Humber and Virginia Almeder.

James M. Humber
Robert F. Almeder

Contents

Contributors

Michael J. Almeida • Division of English, Classics, and Philosophy, University of Texas at San Antonio, San Antonio, Texas

Michael C. Brannigan • Institute for Cross-Cultural Ethics, LaRoche College, Pittsburgh, Pennsylvania

Dennis R. Cooley • History Department, North Dakota State University, Fargo, North Dakota

Susan Dodds • Philosophy Program, Faculty of Arts, University of Wollongong, Wollongong, New South Wales, Australia

Jan C. Heller • Office of Ethics and Theology, Providence Health System, Seattle, Washington

James J. McCartney • Department of Philosophy, Villanova University, Villanova, Pennsylvania

Richard Werner • Department of Philosophy, Hamilton College, Clinton, New York

Abstract

If a human being no longer has a developed capacity for brain activity, we consider it a dead member of the species Homo sapiens.

Although a human corpse is still due some moral consideration, it is not due the full moral consideration we give to a fully fledged person, including a fully fledged right to life. The brain-dead human may have a functioning heart, lungs, and cells. It may be able to sustain itself without machines. Yet, if we have proper permission, once it is pronounced dead we may use its organs for transplant or for scientific research that may, in time, help to ameliorate the human condition.

A nonsentient human embryo has no mental life, for it lacks a brain. It may have a functioning heart, lungs, and cells and may be able to function within the mother's uterus. The nonsentient embryo has the same ontological status as the brain-dead human. Accordingly, it has the same moral status, ceteris paribus.

If the nonsentient human embryo has the same moral status as the dead human being, it is due some moral consideration, but it is not due a fully fledged right to life. Just as we use the organs of a dead human being for transplant or research, given proper permission, so we can use the cells of the nonsentient human embryo for transplant or research, given proper permission, including for stem cell research. They are morally similar cases that should be judged in morally similar ways.

The analogical argument is defended from criticisms based on the potentiality of the unborn, denial that the brain dead human is dead, and the claim that it is wrong to create life to save it.

An Analogical Argument for Stem Cell Research

Richard Werner

The Analogical Argument

Michael S. Gazzaniga, director of the Center for Cognitive Neuroscience at Dartmouth College, argued that in using embryos for research, scientists should regard them the way doctors look upon organs for transplant. When a patient is brain-dead, he said, his organs are harvested. Like the brain-dead patient, Dr. Gazzaniga says, the embryo also lacks a brain.[1]

Like Mr. Gazzaniga, I find the analogy between a brain-dead patient at the end of human life and a not-yet-sentient human embryo or fetus[2] at the beginning of life to be so strong that it is compelling on the issue of stem cell research.

Whether or not a human being has the developed capacity for conscious mental life is determined by whether or not a human being has the developed capacity for brain activity. The developed capacity for brain activity is the *sine qua non*, or indispens-

From: *Biomedical Ethics Reviews: Stem Cell Research*
Edited by: J. M. Humber and R. F. Almeder © Humana Press Inc., Totowa, NJ

able prerequisite, of human life. If we find that the human being no longer has a developed capacity for brain activity, for consciousness, we consider the human being dead. Although a human corpse is still due some moral consideration, that is, it would not do to defile it for fun, it is not due the full moral consideration we give to a fully fledged person. In particular, it is not due a fully fledged right to life. The brain-dead human may have a functioning heart, may be breathing, may have dividing, growing cells, may be able to sustain itself without the assistance of machines. Yet, we still consider it dead when its capacity for mental activity ceases. If we have proper permission, once it is pronounced dead we may use its organs for transplantation into another human or for scientific research. Those transplants or scientific research may, in time, help to ameliorate the human condition.

The brain-dead human is still a human being, a member of the species Homo sapiens. It may be able to function biologically without the assistance of machines, or "naturally," we might say. Its heart may continue to beat. It may continue to breathe. Its cells may continue to divide and grow. Yet we consider it dead once its capacity for mental activity is over. Its death signals the end of its human life and, with it, the end of its fully fledged rights, including the right to life. Its moral status changes with death even though it remains a human being, albeit a dead human being.

Now consider a not-yet-sentient human embryo or fetus. It has no mental life, for it has no brain. It may have a functioning heart and lungs. Its cells may be dividing and growing. It may be able to function without the use of machines within the mother's uterus, or "naturally," we might say. To use the analogy of the brain-dead human, it is not yet a fully fledged person, just as the brain-dead human is no longer a fully fledged person. The not-yet-sentient embryo or fetus has the same ontological status as the brain dead human. Accordingly, it has the same moral status, *ceteris paribus*, that is, assuming all other factors unchanged.

Morally similar cases are to be treated in morally similar ways. Rationality, parity of reason, demands as much. If the nonsentient human embryo or fetus has the same moral status as the dead human being, then it is due some moral consideration; we cannot defile it for fun, but it is not due fully fledged moral consideration. It lacks full moral rights, including a right to life. Just as we can use the organs of a dead human being for transplant or research, given proper permission, so we can use the cells of the not-yet-sentient embryo or fetus for transplant or research, given proper permission, including using its cells for stem cell research. As one goes, so goes the other. They are morally similar cases that should be treated in morally similar ways. That is, in essence, the argument from analogy I find compelling to establish the morality of stem cell research on human embryos.

The Potentiality Counterargument Considered

Arguments from analogy can be slippery. They are only as strong as the analogy itself. Parity of reason only applies if the two items analogized are alike in relevantly similar ways. Morally similar cases are to be treated in morally similar ways just in case they are morally similar in the relevant ways. The argument from analogy for stem cell research just sketched is a first step in an attempt to show that the nonsentient human embryo or fetus is morally similar to a brain-dead human in the relevant ways and, by parity of reason, should be accorded the same moral status and due the same moral consideration.

But what differences can we find between the two that might count as morally relevant differences? One difference is that the presentient embryo or fetus has the potential to become a fully fledged person, whereas a brain-dead human lacks that potential. Indeed, it is just because the brain-dead human lacks the potential for a mental life, lacks a future like ours,[3] that we consider it dead and treat it accordingly. The presentient embryo or fetus may lack the developed capacity for a mental life, but it does have the potential for a mental life, for a future like ours. On the one hand, left in the proper environment, the embryo or fetus will develop naturally and attain a mental life. On the other hand, left in any environment, the brain-dead human will not develop a mental life and have a future like ours, even if its cells continue to divide and grow. Its life is over.

The argument from potentiality attempts to show that the nonsentient embryo or fetus at the beginning of a person's life is not morally akin to the brain-dead human at the end of life. It is argued that the presentient human has the potential for mental activity, including sentience, rationality, and language use that the brain-dead human lacks. Moreover, it is argued that the difference is a morally relevant one because the nonsentient human will naturally grow into a fully fledged person with fully fledged rights whereas the brain-dead human will not. The potential and natural growth of the nonsentient human make it a different moral creature from the brain-dead person and, thereby, connote a different moral status.

To continue the counterargument, the presentient human is due full moral consideration, including a fully fledged right to life, because it is potentially a human adult, whereas the brain-dead human lacks the potential because it is dead. So it is morally permissible to use the organs of the brain-dead human for transplant or research but not morally permissible to do the same with presentient humans. To do so to nonsentient humans is to murder them, to harm them, in a manner not possible with a brain-dead human. It is to deny presentient humans a future like ours.

There are two concepts in this counterargument that carry a great deal of weight: the notion of potentiality and the notion of the natural. I will investigate the notion of potentiality to see whether it can carry the burden of the counterargument.[4]

The argument from potentiality holds that the presentient human has the potential for mental activity and a future like ours and, thereby, should be granted fully fledged moral rights. But the gametes, the ovum and sperm cell, have the same potential for mental activity and a fully fledged human life, including a future like ours. Do human gametes, then, have fully fledged moral rights, including a right to life? Are we obliged, thereby, to see to it that as many gametes as possible are united to continue their natural journey to human adulthood, to their future like ours? If so, then contraception is clearly immoral. So is the use of the rhythm method to avoid pregnancy. Any attempt to avoid pregnancy is an attempt to violate the right to life of the gametes that may be affected. Many, including myself, consider this argument a *reductio ad absurdum*, or reduction to the absurd, of the claim that the potentiality of the nonsentient fetus makes it a morally different creature from the brain-dead human at the end of life.

But the *reductio* is even more striking. If the potential for fully fledged human life gives one full moral rights, including a right to life, then we are obliged not merely to refrain from contraception but to do everything in our power to see to it that as many gametes as possible grow and develop into fully fledged human beings. Just as those who oppose abortion argue that we are obliged to do everything in our power to bring human embryos and fetuses to fulfillment, so by parity of reason, we are obliged to bring all ova to fulfillment, if we take seriously the argument from potentiality. In other words, we are required to fertilize as many ova as possible. If it is wrong to deny to one a future like ours because we ourselves value our own future very highly and morally similar cases are to be treated in morally similarly ways, then the argument should apply to the gametes if it applies to the unborn. If the conclusion that contraception is immoral is not a *reductio ad absurdum* of the counterargument, the conclusion that we are obliged to impregnate every ovum and care for it until birth is.

Moreover, if cloning is possible, we are obligated to clone as many human cells as possible, for any cell that can be cloned into a human is a potential person with a future like ours and, thereby, with fully fledged rights, including a right to life. We may even be obliged to support and develop the techniques of cloning to render them effective, to fulfill the right to life of any cloneable human cell. The parallel here is the manner in which we attempt to save the lives of the unborn and seek better medical technology in the hope that they will survive and attain a future like ours.

It might be objected that the right to life is a negative and not a positive right. Accordingly, one is obliged not to interfere with the life of one who has a right to life but one is not obliged to provide what is necessary for the life to continue. Killing violates a negative right to life but failing to provide what is needed for survival, letting die, does not. Killing is seriously wrong but letting die may not be wrong at all. For instance, it would be wrong to kill those children who will starve to death in underdeveloped nations, even if we kill them painlessly and even if killing them greatly reduces their suffering. But we do not hold it to be wrong to let these same children suffer and die from the effects of hunger, we do it every day.

Using the distinction between killing and letting die, we are obliged not to interfere with the natural life of an ovum, sperm, or cloneable cell, but we are not obliged to provide what it needs for survival. Thereby, we are obliged neither to fertilize every human ovum nor to avoid contraception. Cloning would not be an issue.

This sounds convincing at first. But the use of contraception is interference in the life of the ovum or sperm cell as much as an abortion is interference in the life of the embryo or fetus. In both cases we end a potentially adult person with a future like ours. In both cases we do something, rather than nothing, that prevents a life from continuing, from actualizing its potential. Therefore, a negative right to life does not save the potentiality argument from the conclusion that conception is wrong, from the first *reductio*.

Moreover, we might understand abortion as the right to have the embryo or fetus removed from the mother but not the right to have it killed.[5] The mother's property right over her body gives her the right to have the unborn removed, just as she can have a vagrant removed from her real estate, even if the vagrant was initially invited. But it does not give her the right to kill the unborn or the vagrant even though both could die as a result of the removal. Their death is a collateral consequence of their removal. In both cases, we are treating the right to life and the right to property as negative rights.[6]

If the unborn or the vagrant does die upon removal, it was not killed but allowed to die. The intention was to remove the unborn or the vagrant, not to kill. Death occurs as a collateral consequence of removal.[7] So, if we accept that the right to life is purely negative, not only is stem cell research permitted, so is abortion, both as the removal of the unborn from the mother's property. This is so even if we take seriously the potentiality argument, for the vagrant has not merely the potential to become a fully fledged person, but has actualized the potential, and he or she may still be removed. Reducing the right to life to a purely negative right may avoid complications with gametes but, in conjunction with the potentiality argument, it still entails that contraception is immoral. In addition, it entails that abortion is justified as the removal of the unborn from the mother's property.

Even if the preceding fails, we will now see that the combination of the potentiality argument and reducing the right to life to a negative right would require us to fertilize as many ova or clone as many human cells as possible. Because the unborn is potentially a person, the mother is required to carry it to term. Because the gametes are potentially a person, we are required to see to their fertilization by parity of reason. If human cells are cloneable, we are required to see to their cloning, by parity of reason. As we treat one, so we should treat the other. If we argue that the negative right to life allows us to neglect gametes and cloneable cells and to allow them to die unactualized, then, by parity of reason, the negative right to life should allow the mother to neglect the unborn and allow it to die unactualized, that is, have it removed from her property via abortion. So the move to negative rights saves nothing for the potentiality argument. Even with the move, the potentiality argument falls prey to the same sort of *reductios* we encountered previously.

Perhaps we can save the potentiality argument by arguing that a potential person, like an actual person, must be a single entity. What begins with conception is not biological human life but the life of a specific individual person. To be a potentially fully fledged person, one must be a unique entity that is not disconnected over time and space. Accordingly, the unborn can be a potentially fully fledged person but the gametes cannot. So the *reductios* fail to render unsound the potentiality argument.

This, too, is more complicated than it seems. First, let me distinguish between biological human life and personhood.[8] One is biologically human just as one is a member of the species Homo sapiens. The concept is a biological, scientific one and not a moral concept per se. There is nothing morally compelling about species membership as a biological, scientific concept, *ceteris paribus*. To argue that just because one is a member of a given biological species, one is entitled to full moral consideration is to commit the naturalistic fallacy of the deductivist variety, to infer a normative judgment from a factual statement. Another normative premise is needed in conjunction with the factual, scientific claim about species membership to render such an argument valid. That premise would look something like, "All members of the species *Homo sapiens* have a fully fledged right to life," which, on its face, violates the definist version of the naturalistic fallacy unless further compelling argument can be mounted to establish the truth of the claim. That further compelling argument is not to be found. The question, "*X* is a member of the species Homo sapiens but does *X* have a right to life?" seems a meaningful question to ask, especially when the *X* in question is dead.

Personhood involves sentience as well as more advanced mental capacities, such as consciousness, rationality, self-awareness, and language use. Personhood is a philosophical concept with moral implications because persons, unlike all members of the species of Homo sapiens, can be harmed.[9] As far as we know, only humans are persons. Other species, like apes, whales, and dolphins, may be persons but we do not yet know their status. Aliens from other planets, should they exist, could be persons. Similarly, God, gods, angels, devils, ghosts, and so forth are persons, under most conceptions, if they exist at all. But not all biological humans are persons. Babies and infants are not. Humans with severe mental problems, like acute retardation or insanity, are not. Dead humans are not. Although most humans are persons, not all humans are persons, and it is possible that not all persons are humans.

The argument that we are considering holds that it is necessary that one be a unique individual for one to be a person or a potential person. Because the gametes are not unique individuals, they can neither be persons nor potential persons. Consequently, they cannot serve as counterexamples to the potentiality argument.

But it is not clear that one needs to be a unique individual for one to be a person. Consider those who suffer with multiple personality disorder. They are not unique individuals, yet we still consider them to be persons who have a fully fledged right to life. It is argued by some that each of us is more than one self and that we trade among selves as need be. According to these views, all human selfhood is multiple and not unique.

But, one might counter, the previous paragraph refers to mental aspects of the self and not to physical aspects. A person must be a unique physical being even if she is not a unique being psychologically.

This, too, is problematic.

But this claim can also be disputed on empirical grounds. It is not clear that the zygote is the same organism or proto-organism as the embryo that may later develop from it. During the first few days of its existence, the conceptus subdivides into a set of virtually identical cells, each of which is "totipotent"—capable of giving rise to an embryo. Spontaneous division of the conceptus during this period can lead to the birth of genetically identical twins or triplets. Moreover, it is thought that two originally distinct zygotes sometimes merge, giving rise to a single and otherwise normal embryo. These facts lead some bioethicists to conclude that there is no individuated human organism prior to about fourteen days after fertilization, when the "primitive streak" that will become the spinal cord of the embryo begins to form.[10]

So the notion of a unique individual does not save the argument from potentiality even as amended for a potential person. The uniqueness of the organism is not clear at conception and not for at least 2 weeks into its development. Consequently, it would fail to be a potential person possessing a fully fledged right to life. As such, the embryo could be harvested for stem cells before the primitive streak occurs.

Furthermore, it is not obvious that a person must be a unique physical being. Indeed, insofar as God, gods, angels, devils, ghosts, and so on are conceivable as persons, they lack a physical being altogether, at least under most descriptions. Under some descriptions, God is considered to be omnipresent, in all places at once but at no specific place at any moment. We can imagine aliens who have disjointed bodies but are considered persons. They might, for instance, identify with the various disjoints, care for the disjoints, have the experiences of the disjoints, control the movement of the disjoints, and so on. *Star Trek's* Borgs approach such a condition when the Borg hive behaves as a single person but is made of many physically disconnected Borgs or parts. We might consider the Borg collective a single person with many disjoints. Conceptually, a person need not be a unique physical being. It may be that members of the species *Homo sapiens* need be physically unique but that has nothing to do with the potentiality argument. The unborn are human. The issue is whether their potential personhood, their potential future like ours, grants them a right to life. As we have seen, not all humans have a right to life, especially not dead humans.

The argument from potentiality holds that the nonsentient human has the potential for mental activity and a fully fledged human life and, thereby, should be granted fully fledged moral rights. But, one might ask, why should it be accorded fully fledged rights, including a right to life? At age 8, I was a potential motorist, a potential property owner, and potentially morally autonomous. But, at age 8, my potential for these did not accord me the right to them. At age 8, I had the potential rights to drive a car, own property, and formulate my own moral judgments, but those potential rights did not grant me the actual rights in question. They were potential rights and not actual rights. Similarly, the potential person may have potential rights, like a potential fully fledged right to life, but that does not convey an actual right and, in particular, it does not convey the claim against others that an actual right ensures. Simply put, a potential right is not an actual right, as the previous examples make clear. Just how, then, does the potential person have fully fledged rights, including a right to life? How is its potentiality a license for actual rights? The argument of this paragraph, if successful, is itself sufficient to refute the potentiality argument.[11] All in all, I conclude that the concept of potentiality cannot be used to discredit the analogy between the end-of-life, brain-dead human and the nonsentient fetus without itself creating a series of defeating *reductio ad absurdum* arguments or falling into conundrums explaining how the potential for personhood grants actual rights.

"Denial That the Brain-Dead Human is Dead" Considered

Suppose those who argue that stem cell research is wrong are willing to accept the analogy between the brain-dead human and the nonsentient human embryo. Now suppose they run the analogy the other way. That is, because it is seriously wrong to kill the nonsentient human embryo, it is seriously wrong to consider the brain-dead human as a dead human and to kill the brain-dead human. In other words, suppose that the defender of the prohibition on stem cell research argues that using brain death as a criterion for human death is a mistake. The defender of the prohibition on stem cell research might offer some other criterion for death, like the end of spontaneous heartbeat or the end of spontaneous respiration or the end of cell growth.

The problem with this move is that it is costly. People can go on living in this new sense long after their developed capacity for mental activity ceases. Usually they must be kept alive by machines. The use of the machines is costly. Even if they do not require machines, they must be feed intravenously and their needs attended. Expenses are involved. Who is responsible for the expenses?

The obvious response is that the dying person (dying by our new definition but dead by the old) or the closest relatives of the dying person are responsible. But suppose they refuse to pay— then, the services stop and the dying person is allowed to die, not killed but allowed to die. This is no different from present procedure. We have merely changed the criterion of death but letting die goes on as before. Death just occurs at a different time.

Now consider the nonsentient embryo or fetus. The mother refuses to pay, to provide the property necessary for the maintenance of the nonsentient embryo or fetus. She removes her property just as the dying people or the relatives remove theirs by refusing to pay. The mother is removed from the embryo or fetus just as the life support is removed from the dying person. Just as we may use the now dead person's organs for transplant,[12] given proper permission, so we may use the stem cells of the embryo or fetus for research, given proper permission. Changing the criterion for death does not help the opponent of stem cell research.

There are other problems. What now will count as the death of the embryo or terminally ill patient? Which criterion for death do we adopt? Must all cell division cease? If so, we may have to wait a very long time for the terminally ill person to die. Hair and fingernails continue to grow for quite some time after all of the other cells are dead. By that time, no organs would be viable for transplant nor would the embryo be capable of supplying viable stem cells for research. More to the point, it would be a very expensive and messy affair with all of those not-quite-corpses decomposing but their hair and fingernails in need of a trim. If hair and fingernail cells do not count, which cells do count? In a nonquestion-begging way, why do just those and not others count?

Perhaps we could use the cessation of spontaneous heartbeat or respiration as the criterion for death. The problem is that some who still have a developed capacity for mental activity no longer have spontaneous heartbeat or respiration. The new criterion would allow us to kill many we now consider fully fledged persons. Many who would otherwise be saved to live out a normal life would, by the decree of the new criterion, be left to die. There would be nothing wrong with doing so, if we change the criterion for death as suggested. Yet it seems wrong for all of the reasons we now believe that brain death is the proper criterion for death.

Moreover, because the human embryo has neither a heart nor lungs, it would not be covered by either criterion. We have already seen that its potential for a heart or lungs would not help because the gametes have the same potential. So, by these two criteria, we should be able to pursue stem cell research. The very rationale for opting for the new criteria collapses.

"It is Wrong to Create Life in Order to Destroy It" Considered

Perhaps the most popular argument against stem cell research is that it is seriously wrong to create life to destroy it. Taken literally, this is absurd. We create life to destroy it when we farm food, be it fruit, vegetable, or animal. So, I take it, the popular claim is really, "It is wrong to create *human* life to destroy it."

This, too, is false on its face. We create human life to destroy it whenever there is intentional and consensual human conception. We know that any human who is conceived will be destroyed. Everyone dies. Perhaps the difference here is that it is not the parent who will do the destroying but it is the medical researcher who destroys the embryo. Normally, one is allowed to die on one's own terms. Normally, one is not destroyed or killed.

As the lyrics to the folk song "Where Have All of the Flowers Gone?" make clear, mothers have given birth to children only to send them to die in war for most of human civilization. Mothers still do. Is this creating life to destroy it? Perhaps one can retort that again we are involved with the letting die of soldiers and not the killing of soldiers. The enemy does the killing. We merely send them off to die. So we don't destroy them, the enemy does. But there is another problem. It is seriously wrong to destroy human life, all other things being equal. So the following statement is true, "It is seriously wrong to jump rope to destroy human life, all other things being equal." So is the statement, "It is seriously wrong to chew gum to destroy human life, all other things being equal." Indeed any general statement is true, if it has the form, "It is seriously wrong to X to destroy human life, all other things being equal." Any activity that is undertaken to cause a serious wrong is itself seriously wrong, all other things being equal. So what is really at issue is the destruction of human life and not creating it to destroy it. Common wisdom holds that the destruction of human life is justified in some cases: self-defense, a just war, or capital punishment to mention a few. It is this realization that is covered by the *ceteris paribus* clauses above. So the prohibition against destroying life is not absolute. It is a *prima facie*, a conditional wrong not an absolute wrong. Are there other instances when we think we are justified destroying a human?

We have already mentioned one. We think we are justified destroying a dead human. We may embalm, cremate, or bury a dead human. In some cultures, we may feed it to birds. In our culture we may use a dead human as a source of organ transplant given the proper permission. So, by parity of reason, we ought to think we are justified destroying a not-yet-alive human. Surely a dead human is not a live human even if its cells are dividing and growing, its heart is beating, or its lungs are working. So long as the dead human no longer has a developed capacity for brain activity, it is dead and no longer alive.

So before the development of the capacity for brain activity, for sentience, for consciousness, the unborn is a not-yet-alive human. Like the dead human, it is human. Like the dead human, it is not alive. Because we can destroy the dead human given the proper permission, so we can destroy the not-yet-alive human given the proper permission. Because sentience does not appear until sometime in the second or third trimester,[13] and the embryo stage ends long before this time, the creation of embryos for stem cell research is a morally legitimate activity, for the embryo is not yet a living human being.

My point here is that humans are different from most other creatures. Because an adult human is normally a person, it is not enough merely to have cell division or blood circulation or respiration to be considered a living human being. Humans are different because they normally produce persons as their adult members, persons who are conscious, can reason, use language, are self-aware, are members of moral communities, and so forth. Because an adult human being is normally a person, a human being must be sentient to be a living human. It is our mental life that is the *sine qua non* of personhood. That mental life does not begin until sentience or consciousness begins. It is reasonable to hold that we are not alive until we are sentient, conscious. We embrace the view at the end of human life and, by parity of reason, we should embrace it at the beginning of human life. Whether at the beginning or the end of life, no developed capacity for mental activity, no living human being. Given proper permission, we can destroy dead humans. For all of the same reasons, given proper permission, we can destroy presentient humans, that is, nonliving human embryos or fetuses. Given proper permission, we can use the organs of brain-dead humans for transplant. Given proper permission we should be able to use the cells of nonsentient embryos or fetuses for stem cell research. Human life begins with sentience and not with conception. If it did begin with conception we would be hard-pressed to explain why we can destroy the dead even with the proper permission.

Nevertheless, it does not follow that we can destroy nonsentient human embryos or fetuses for frivolous reasons. We should respect one another's attributions of moral status whenever we can do so without sacrificing anything of greater moral importance.[14] Many people attribute full moral status to the nonsentient embryo or fetus. All other things being equal, we should honor their attributions. But we are not required to honor their attributions when doing so involves a greater moral sacrifice. This is especially so, given that their attributions are not fully tenable. Judges are allowed to override a family's decision to deny an autopsy, if there are good legal reasons to do so, reasons that serve the interests of legal justice. In that case, the family's attribution of moral status to the dead is not fully tenable from a legal perspective. Similarly, scientists should be allowed to do research on embryonic stem cells, if they have the proper permission and regardless of people's attribution of moral status, if there is good reason to believe that that research can advance medical science and either save actual human lives or alleviate actual human suffering. Although we have a duty to honor the moral attributions of others, it is a weak moral duty that can be overridden by other weightier moral considerations like the attempt to ameliorate the human condition.

So, I conclude that nonsentient embryos and fetuses are just the sort of humans we can destroy, given the proper permission, because they are not yet alive. We should treat them like brain-dead humans for they are, in all morally relevant respects, like brain-dead humans.

A Conceptual Challenge to Analogical Arguments Considered

I will end by considering a conceptual challenge to the argument from analogy. There are those who are skeptical of arguments from analogy like the one that I offer. They consider them mere intuition-pumps and reason that if one is clever enough then there is always an intuition-pump to prove the contradictory position. There is no reason to believe that our intuitions are consistent nor, by extension, our intuition-pumps. They may be right, although their case would be more compelling if they could produce the intuition-pump on the other side. I do not know of any for the case under consideration. For those who are so bothered, however, I recommend Ernle' W. D. Young's article, "Ethical Issues: A Secular Perspective."[15] Using moral principles developed and defended by Mary Anne Warren in the book *Moral Status*,[16] Young derives the same conclusion on stem cell research that I do from analogy. Because I admire both the moral principles and Warren's work, I find the derivation felicitous. In addition, I believe my position to be compatible with that taken by David Boonin in *A Defense of Abortion*.[17]

Acknowledgments

I thank John Bartle for his help on an earlier draft of this work.

Notes and References

[1]*The New York Times on The Web*, January 19, 2002, "Science Academy Supports Cloning To Treat Disease," by Sheryl Gay Stolberg, page 3. Warren, M. A. (1997) *Moral Status*. Oxford University Press, New York, offers a similar argument.

[2] Although I will use the phrase "embryo or fetus" throughout, I mean by it the unborn human from conception to birth.

[3]Marquis, D. (1989), Why abortion is immoral. *J. Philos* **April,** 183–202, uses the phrase "future-like-ours" to describe the unborn's future and conjoins that with the claim that one's future is among those things most valuable to one and to deny one of one's future is a serious wrong. But it seems clear that the unborn's future is of no value to the unborn because it is not capable of valuing anything. To assume that it would be of value to the unborn if it were like us is to beg the very question at issue, that is, to assume that the unborn is enough like us to be capable of conscious states like desire, which it is not. Nor is it clear that universalizability commits me to valuing the unborn's future because I value my own. Requiring universalizability for the unborn also begs the question of its status since we only need to universalize for those who are enough like us to be due serious moral concern. In addition, the gametes' future are as much a value to them as the unborn's future is to it. For a comprehensive critique of Marquis' position, *see* Boonin (2002) *A Defense of Abortion.* Cambridge University Press, New York, pp. 56–85.

[4]For compelling arguments that the notion of the natural is deeply problematic as a moral concept, *see* Holmes, R. L. (2003) *Basic Moral Philosophy* 3rd ed. Wadsworth Publishing, Belmont, CA, pp. 102–104, and Leiser, B. M. (1979) *Liberty, Justice and Morals: Contemporary Value Conflicts,* 2nd ed. Macmillan, New York, pp. 52–59. For compelling arguments that the notion of potentiality is not helpful for establishing a right to life in nonsentient embryos and fetuses, *see* McMahan, J. (2002) *The Ethics of Killing.* Oxford University Press, New York, especially pp. 302–329; David Boonin, *op. cit.,* especially pp. 45–49 & 56–90; Warren, *op. cit.,* especially pp. 205–208.

[5]Thomson, J. J. (1971) A defense of abortion, *Philosophy and Public Affairs* **1(1),** 47–66, defends the view that abortion is the right to have the unborn removed but not the right to have it killed. Present practices do involve killing the fetus *in utero* but the practices could be changed to require removing the fetus and allowing it to die *ex utero.* I am not recommending the practice or the view but merely reporting on them.

[6]Humber, J. (1975) Abortion: the avoidable moral dilemma. *The Jour-*

nal of Value Inquiry, **9(4)** 282–302 offers an interesting twist on Thomson's reasoning. Those who oppose abortion would be well served if they put their resources into developing technological means of saving the aborted and bringing them to infancy just as we might provide means to save the homeless person removed from someone's property.

[7]McIntyre, A. (2001) Doing away with double effect. *Ethics* **111(2)**, 219–255, provides an excellent critique of the principle of double effect, a principle conjoined with the potentiality argument in the reasoning of many who oppose abortion and those who hold Thomson's position. Like McIntyre, I find the principle unacceptable. But for the sake of the argument I allow it to stand to examine the potentiality argument. As such, my critique of the potentiality principle plus double effect, the standard conservative position on abortion or stem cell research, is an internal rather than an external one. Like many philosophers, I am suspicious of the positive/negative right and duty distinction. Shue, H. (1997) *Basic Rights*, 2nd ed. Princeton University Press, Princeton, provides an excellent critique of the supposed distinction. Again, I allow the distinction to stand for the sake of the argument.

[8]Warren, *op. cit.*, pp. 18–19

[9]Dead humans cannot be directly harmed nor can nonsentient humans. Neither has the developed capacity for harm, that is, consciousness. *See* Warren, *op. cit.* pp. 204–205. Boonin, *op. cit.*, has a particularly acute discussion of when the relevant sort of consciousness occurs placing it sometime after 25 weeks of pregnancy, *see* pp. 98–129.

[10]Warren, *op. cit.*, pp. 205–206.

[11]Wilkins, B. T. (1993) Does the fetus have a right to life? *Journal of Social Philosophy* **XXIV(1)**, 123–137, questions the line of counterargument offered here. Boonin, *op. cit.*, pp. 45–48, effectively responds to Wilkins.

[12]This may be difficult depending on the criterion of death. The organs are more likely to be unusable, in some cases, if we wait for the end of spontaneous heartbeat or respiration. They will surely be unusable if we wait for the end of all cell division and growth.

[13]Boonin, *op. cit.*, pp. 98–129; *see also* Warren, *op. cit.*, pp. 204–205.

[14]Warren, *op. cit.*, pp. 170–172.
[15]Holland, S., Lebacqz, K., and Zoloth, L., eds. (2001) *The Embryonic Stem Cell Debate*. MIT Press, Cambride, MA, pp. 163–174.
[16]Warren, *op. cit.*
[17]Boonin, *op. cit.*

Abstract

The inclusiveness thesis is now widely endorsed among moral philosophers and other members of the moral community. The thesis claims that the moral community includes the set of all sentient beings. The developmental disadvantages of those human beings who lack rationality or a sense of the past and future or autonomy or imagination, and so on, do not exclude these beings from direct moral standing. The most powerful argument for the inclusiveness thesis is the well-known argument from marginal cases. I show first that the argument from marginal cases is unsound and therefore cannot establish the inclusiveness thesis. I offer instead the impartial argument from marginal cases and show that the impartial argument from marginal cases establishes both the inclusiveness thesis and the greater inclusiveness thesis: the position that early-term fetuses and human embryos have direct moral standing. I conclude that early-term fetuses and human embryos have all of the protections extended to full members of the moral community. Lethal experimentation on early-term fetuses and human embryos therefore demands the sort of direct moral reasons that would justify lethal experimentation on more developed members of the moral community.

Marginal Cases and the Moral Status of Embryos

Michael J. Almeida

Introduction

It is widely held among moral philosophers that membership in the moral community guarantees direct moral standing to each member. Direct moral standing ensures that the life and well being of each member matters in a special way: members of the moral community are assured protection from, among other things, cruelty, mistreatment, neglect, destruction, and exploitation. As social contract theorists have long emphasized, there are very strong prudential reasons to gain entrance into the moral community.

It is also widely held that no special protection extends to those outside the moral community. Those who are often excluded from the moral community—cows, pigs, lambs, goats, early-term fetuses, human embryos, and so on—are therefore at a very serious disadvantage. The destruction of those without direct moral standing does not in general require any moral justification at all.

From: *Biomedical Ethics Reviews: Stem Cell Research*
Edited by: J. M. Humber and R. F. Almeder © Humana Press Inc., Totowa, NJ

Recent work in moral theory bodes well for many beings traditionally excluded from the moral community. It is now a dominant view among moral theorists that the moral community extends to all sentient beings. One may call the thesis that every sentient being has direct moral standing the *inclusiveness thesis*. The most powerful argument for the inclusiveness thesis is the argument from marginal cases (AMC). AMC challenges human beings to provide a principled and relevant reason to include *all* and *only* human beings in the moral community. Every well-known criterion for admission to the moral community—rationality, language ability, self-consciousness, a sense of the future, memory, an emotional life, imagination, purposiveness, and so on—excludes at least some so-called marginal human beings. Because predominant moral judgment rejects the proposal that marginal human beings have no direct moral standing, we are moved to conclude that sentience is the criterion for direct moral standing. The moral community is therefore expanded to include all sentient beings.

Of course, the inclusiveness thesis does not include every marginal human being in the moral community. It remains a dominant view among moral theorists that the moral community does not extend beyond the set of sentient beings. One may call the thesis that there is direct moral standing for early-term human fetuses and human embryos the *greater inclusiveness thesis*. Most moral theorists maintain that the greater inclusiveness thesis is false. Nonsentient human beings are not among those included in the moral community.

In the next section, *The Argument From Marginal Cases*, I show that the standard argument from AMC cannot establish the inclusiveness thesis. AMC cannot provide a reason acceptable to everyone that sentience is a sufficient condition for direct moral standing. In the following section, *The Impartial Argument From Marginal Cases*, I show that an improved argument from marginal cases—the impartial argument from marginal cases or IAMC—can provide a reason acceptable to everyone that sentience is a sufficient condition for direct moral standing. IAMC, therefore, establishes the inclusiveness thesis. But IAMC also provides a reason acceptable to everyone that early-term fetuses and human embryos are members of the moral community. IAMC therefore establishes the controversial greater inclusiveness thesis. I conclude that the inclusiveness thesis is true only if the greater inclusiveness thesis is true. We are therefore led to conclude that early-term fetuses and human embryos have direct moral standing.

The greater inclusiveness thesis entails that there is no greater moral justification to destroy human embryos in stem cell research than there is to destroy other members of the moral community—including more developed sentient, rational, or self-conscious beings—in research demanding the extraction of stem cells. All of these beings are members of the moral community and all have direct moral standing.

The Argument From Marginal Cases

AMC is designed to expose a disposition toward *speciesism* among human beings. Speciesism is the unprincipled and morally unjustified partiality toward one's own species. Because there is no significant difference between some human beings and sentient nonhumans, there is no principled basis for including every human being in the moral community and excluding every sentient nonhuman. Below is a standard version of the argument from marginal cases.

1. Many sentient human beings—including human infants, the profoundly developmentally challenged, and others— lack the normal adult qualities of rationality, language use, purposiveness, self-consciousness, memory, imagination, expectation, a sense of the past and future, and so on. These are the so-called marginal human beings.
2. There are many other species that lack the qualities found in normal adult human beings but are nonetheless capable of enjoying pleasurable experiences and suffering painful

experiences. These are the so-called sentient nonhumans.

3. Marginal human beings rightly have direct moral standing as members of the moral community. There are direct moral objections to raising marginal human beings for food, subjecting them to lethal scientific experiments, treating them as chattel, and so on.

4. Sentient nonhumans do not rightly have direct moral standing and are not included in the moral community. There are no direct moral objections to raising sentient nonhumans for food, subjecting them to lethal scientific experiments, treating them as chattel, and so on. (Assumption)

5. If marginal human beings rightly have direct moral standing and sentient nonhumans do not rightly have direct moral standing, then there must be some morally significant property R that marginal humans possess and sentient nonhumans lack.

6. There is no morally significant property R that marginal human beings possess and sentient nonhumans lack.

7. Therefore, either marginal human beings do not rightly have direct moral standing or sentient nonhumans do rightly have direct moral standing (from premises 5 and 6).

8. Therefore, sentient nonhumans rightly have direct moral standing (from premises 3 and 7).[1]

The argument from marginal cases is clearly valid, but there is good reason to doubt its soundness. Consider, for instance, premise 6. Among the proposals for relevant property R, some have received almost no serious consideration. The property "made in the image and likeness of God" or the property "ensouled beings who are the objects of salvation" at least appear to be morally relevant properties that characterize all and only human beings. But it is generally taken to be a decisive objection to these proposals that the possession of such properties is *unverifiable*. Lawrence Becker offers a typical observation.

. . .[T]here does not seem to be a morally relevant characteristic that distinguishes all humans from all other animals.
Sentience, rationality, personhood, and so forth all fail. The
relevant theological doctrines are correctly regarded as
unverifiable and hence unacceptable as a basis for a philosophical morality.[2]

But the violation of verifiability is by no means decisive,
contrary to what Becker and others seem to suggest. There simply is no convincing argument that verifiability constitutes a suit-
able criterion of meaning, existence, or anything else. Indeed, no
version of verifiability has been credible since the lost days of
logical positivism. So we are in no position to conclude that only
properties whose possession we can verify—in some sense or
other of "verify"—matter morally. And so we cannot conclude
that there does not exist a morally relevant property that all and
only human beings possess.

But the more serious objection to the soundness of the argument from marginal cases concerns premise 3. There is no rationale in AMC for the claim that marginal human beings should
have direct moral standing as members of the moral community.
It is no doubt abhorrent to most human beings to consider raising
marginal humans for food or treating them as chattel. It is also
true that predominant moral judgment prohibits using marginal
humans in scientific experimentation even for the most admirable
goals. But moral judgment and moral sensibility, however widely
shared, do not constitute good moral reason for including marginal humans in the moral community. After all, it is also a predominant moral judgment that sentient nonhumans are *not*
members of the moral community. However, that popular moral
judgment is regarded by many as poor reason for denying sentient nonhumans direct moral standing. Therefore, dominant
moral judgments, even well-considered moral judgments, are
simply not sufficient to establish premise 3.

Imagine a community consisting of beings that possess the standard list of possibly relevant capacities. Let's call this group the *advanced community*. Each member of the advanced community has the properties of rationality, language use, purposiveness, self-consciousness, memory, imagination, expectation, a sense of the past and future, and so on. In addition to the standard list of possibly relevant abilities each member of the advanced community possesses a capacity for telepathy, psychokinesis, psychoprojection, mental healing, and so on. A whole series of possibly relevant mental capacities that normal adult human beings lack. Members of the advanced community might arrive at the conclusion that R specifies a cluster of morally relevant properties, including many that normal adult human beings do not possess. Now suppose that there are no marginal cases in this community and, as a matter of contingent fact even the least developed members of the advanced community possess capacities sufficient for inclusion in the moral community. Members of this community might quickly arrive at the conclusion that direct moral standing does not extend to every adult human being and certainly not to every sentient nonhuman.

The advanced community offers a principled basis for excluding every sentient nonhuman and every human being from the moral community. The standard argument from marginal cases therefore offers them no reason at all to accept the inclusiveness thesis. Of course there are familiar highhanded responses available. We might simply insist, for instance, that normal adult human beings are members of the moral community and discontinue the discussion. Or we might charge the advanced community with an underdeveloped moral sense and end discussion in this way. But these highhanded responses do not constitute a reason for anyone in the advanced community to accept the inclusiveness thesis. Indeed it is difficult to escape the conclusion that these responses simply beg the question in our favor.[3] But do we really have no reason acceptable to everyone—including everyone in the advanced community—that every human being is rightly included in the moral community?

The Impartial Argument From Marginal Cases

Certainly everyone will agree that the advanced community might be unfairly favoring itself. The argument against including normal adult human beings in the moral community appears to take advantage of certain contingent characteristics of the advanced community. For instance, members of the advanced community enjoy, as a matter of contingent fact, certain developmental advantages over other species and other beings. If speciesism is the unprincipled partiality toward one's own species, then we should add that *developmentalism* is the unprincipled partiality toward one's own level of development. But how do we show that those developmental advantages do not constitute a morally relevant difference between the advanced community and the human community?

We might suppose that no member of the advanced community or human community knows the level of development they happen to enjoy. If the possession of certain developmental advantages constitutes a morally relevant difference between the advanced community and the human community, then the possession of such advantages would still constitute a morally relevant difference were it unknown to everyone who possessed them. But if it is unknown to everyone who possesses those developmental advantages then no member of either community is in a position to unfairly exploit their possession. Members of these communities are therefore in a position to consider whether an impartial assessment of these developmental advantages would lead them to the conclusion that these advantages constitute a morally relevant difference.

But it is very unlikely that any member of the human community or the advanced community would risk exclusion from the moral community on the chance possession of certain contingent properties, psycho-kinetic properties for instance, or psycho-projection, or mental healing, or even rationality, and so on. We are simply not that sure that any one of these properties makes a morally relevant difference. And membership in the moral community provides protection for each member that is too valuable to risk on the possession contingent properties that might be morally relevant. And so we will not find members of these communities converging on the conclusion that these developmental advantages constitute a morally relevant difference between the human community and the advanced community.

The argument against including normal adult human beings in the moral community takes advantage of certain contingent characteristics of the advanced community. But those developmental advantages do not provide a morally relevant reason for favoring anyone. Excluding adult human beings from the moral community because adult humans lack such advantages is therefore a form of developmentalism. It is simply an unprincipled partiality toward the level of development found in the advanced community.

Notice, however, that an impartial consideration of the contingent advantages of adult human beings also provides no moral reason for excluding sentient nonhumans from the moral community. It is again unlikely that any member of the human community would risk exclusion from the moral community on the chance possession of rationality, language use, purposiveness, self-consciousness, memory, imagination, expectation, a sense of the past and future, and so on. No one is sufficiently certain that rationality, language use, and so on, are properties that make a moral difference large enough to exclude sentient nonhumans from the moral community. And so the impartial argument from marginal cases also provides a convincing basis for the inclusiveness thesis.

But among those who accept the inclusiveness thesis, the developmental advantages of normal adult human beings are widely regarded as a relevant reason for excluding early-term fetuses and human embryos. It is obvious that normal adult human beings enjoy many developmental advantages over early-term fetuses and human embryos. Early-term fetuses and human embryos are not rational or purposive, for instance, and they have no sense of the past and future, and so on. But let's consider how certain we are that these developmental advantages constitute a moral difference large enough to exclude early-term fetuses and human embryos from the moral community.

Suppose that no member of the human community knows the developmental point that—as a matter of contingent fact—he or she happens to have reached. If early-term fetuses and human embryos are not members of the moral community, then they possess no moral protection from destruction and exploitation. No direct moral justification is required for the termination of early-term fetuses and human embryos.[4] But human beings who are prepared to permit the termination of human embryos or early-term fetuses must be prepared to prevent the existence of many adult human beings. Of course we have assumed that no one knows his or her own level of development—no one knows whether he or she is a normal adult human being, a newborn, an early-term fetus, or not yet existing.[5] So everyone is aware that, for all anyone knows, his or her own precursor currently exists among those early-term fetuses and human embryos. Those prepared to permit the termination and exploitation of human embryos must therefore be ready to forfeit a world in which he or she exists as a normal adult human being for a world in which he or she never existed or never existed as an adult.[6] But certainly no one is willing to risk his or her normal adult existence on the chance occurrence that he or she has already reached a normal adult level of development. No one is *that* certain that the developmental advantages of normal adult human beings make a moral difference so significant that the termination of his own precursor is morally unimportant.[7] And there is obviously a great deal to lose if, as we found in our discussion of the advanced community, those developmental advantages make no moral difference at all.[8] For those who are not in a position to exploit their own level of development, then, it is not unreasonable to conclude that the moral community includes early-term fetuses and human embryos. The impartial argument from marginal cases therefore offers a convincing basis for the greater inclusiveness thesis.

Human beings who insist that there are developmental reasons for excluding early-term fetuses and human embryos from the moral community are therefore developmentalists. The developmental reasons advanced for excluding human embryos from the moral community are not reasons anyone would accept for excluding their own precursors from the moral community. Excluding early-term fetuses and human embryos from the moral community simply expresses an unprincipled partiality toward the level of development found in normal adult human beings.

Concluding Remarks

Every member of the moral community has direct moral standing. Direct moral standing provides protection from, among other things, cruelty, mistreatment, neglect, destruction and exploitation. The inclusiveness thesis claims that the moral community includes among other beings every sentient nonhuman. The greater inclusiveness thesis claims that the moral community includes among other beings early-term fetuses and human embryos. We found that the impartial argument from marginal cases establishes both the inclusiveness thesis and the greater inclusiveness thesis. We appealed to the following impartial argument from marginal cases, now presented in standard form:

1*. Normal adult human beings rightly have direct moral standing as members of the moral community. There are direct moral objections to subjecting normal adult human beings to lethal scientific experiments, even for good goals.
2*. Early-term fetuses and human embryos lack the developmental advantages of normal adult human beings, including rationality, language use, purposiveness, self-consciousness, memory, imagination, expectation, a sense of the past and future, and even sentience.

3*. Early-term fetuses and human embryos do not rightly have direct moral standing and are not included in the moral community. There are no direct moral objections to subjecting them to lethal scientific experiments (Assumption).

4*. If normal adult human beings rightly have direct moral standing and early-term fetuses and human embryos do not rightly have direct moral standing, then there must be some morally significant property, termed R, that adult human beings possess and early-term fetuses and human embryos lack.

5*. There is no morally significant property R that adult human beings possess and early-term fetuses and human embryos lack. The developmental advantages of adult human beings do not constitute a morally relevant difference between human embryos and normal adult human beings.

6*. Therefore either adult human beings do not rightly have direct moral standing or early-term fetuses and human embryos do rightly have direct moral standing (from premises 4* and 5*).

7*. Therefore, early-term fetuses and human embryos rightly have direct moral standing (from premises 1* and 6*).

The impartial argument from marginal cases is sound if premise (5*) is true. But we established in the previous section that it is reasonable to believe that premise (5*) is true. Perhaps more cautiously we found that no one would risk his normal adult existence on (5*) being false. And this is because we are simply not that certain that (5*) is false. For similar reasons members of the advanced community were unwilling to risk membership in the moral community on the moral relevance of psychokinetic powers, or telepathic powers, or powers of mental healing. No one is so certain that these are relevant capacities that they are willing to risk their own exclusion from the moral community on it.

Among the central consequences of the impartial argument from marginal cases is that lethal experimentation on early-term fetuses and human embryos is morally unjustified in the absence of direct moral reasons. The developmental differences between adult human beings and human embryos are not sufficient to exclude human embryos from membership in the moral community. Therefore, the lives of early-term fetuses and human embryos have every protection afforded members of the moral community. It is well known that stem cell research on early-term fetuses and human embryos has extraordinarily promising implications for medical therapy.[9] But if early-term fetuses and human embryos are members of the moral community—and we have offered reasons to believe that they are—then the termination of fetuses and human embryos in well-intended research lacks all of the moral justification of the termination of more advanced humans in well-intended research.

Notes and References

[1]The foregoing version of AMC is based on Lawrence Becker's formulation in Drombroski, D. A. (1997) *Babies and Beasts: The Argument from Marginal Cases*. University of Illinois Press, Chicago. *See* Drombroski for an extended discussion of the importance of AMC in recent moral debate.

[2]*See* Becker, L. (1983) The priority of human interests, in *Ethics and Animals* (Miller, H. and Williams, W., eds.), Humana Press, Totowa, NJ.

[3]*See*, for instance, Kuhse, H. and Singer, P. (2002) Individuals, humans, and persons: the issue of moral status, in *Unsanctifying Human Life* (Kuhse, H., ed.), Blackwell Publishers, Oxford.

Why do we think that killing human beings is so much more serious than killing these other [sentient non-human] beings? . . . The *obvious* answer is that human beings are different from other animals, and the greater seriousness of killing them is a result of these differences. But which of the many differences between human beings and other animals justify such a distinction? Again, the *obvious* response is that

the morally relevant differences are those based on our supe-
rior mental powers—our self-awareness, our rationality, our
moral sense, our autonomy, or some combination of these... .
That the particular objection to killing human beings rests on
such qualities is very plausible (p. 193, *my emphasis*).

But these responses are not at all obvious and simply beg the question
against the advanced community, who find such qualities of little
or no moral significance.

[4]Of course, the termination of human embryos might be prohibited for
its instrumental disvalue. But because human embryos are not
members of the moral community, the termination *per se* of human
embryos is not something for which a moral justification is required.

[5]I do not assume in this argument that anyone is identical with the
embryo or fetus that is his precursor. On some theories of personal identity that identification fails, on others it does not. But
neither assumption affects the argument that follows. The argument also does not assume that the embryo possesses (or is identical to) a hylomorphic soul or a Cartesian soul. For an interesting
series of arguments on whether the embryo is ensouled and the
permissibility of killing "early life," *see* McMahon, J. (2002) *The
Ethics of Killing.* Oxford University Press, Oxford, pp. 7–19 and
p. 267 ff.

[6]The argument I am proposing applies to existing human beings at
various levels of development. I do not conclude or propose that
potential human beings—unfertilized ova, for instance—are
also members of the moral community. But it should be clear
that I am not using "human being" synonymously with "human
person."

[7]But compare Kuhse, H. and Singer, P. (2002) The moral status of the
embryo, in *Unsanctifying Human Life* (Kuhse, H., ed.) Blackwell
Publishers, Oxford. In a series of hypothetical cases Singer contends that there is no obligation to preserve the lives of embryos.
Suppose it is permissible to dispose of an egg and sperm from a
certain couple. If so, then argues Singer, suppose

... .[T]he couple is asked if they are prepared to consent to
the newly created embryo being frozen to be implanted into
someone else, but they are adamant that they do not want

their genetic material to become someone else's child. Nor is there any prospect of the woman's condition [viz. her medical condition precluding the possibility of pregnancy] ever changing, so there is no point in freezing the embryo in the hope of reimplanting it in her at a later date. The couple ask that the embryo be disposed of as soon as possible. (cf. pp. 182–183)

Now Singer asks "how plausible is the belief that it was not wrong to dispose of the egg and sperm separately but would be wrong to dispose of them after they have become united?"
I take it that the question is rhetorical. But the impartial argument from marginal cases would have us conclude that the embryo is a human being (the sperm and egg separately are only potential human beings) and a member of the moral community.
[8] We have arrived at the conclusion that the developmental differences between adult human beings and human embryos is not so significant that human embryos are excluded from the moral community. The reasoning can be set out in matrix form.

	Qualities Are Relevant	Qualities Are Not Relevant
I Conclude Qualities Are Relevant	Loss of My Adult Life (Permissible)	Loss of My Adult Life (Extreme Moral Disvalue)
I Conclude Qualities Are Not Relevant	No Loss of My Adult Life (Permissible)	No Loss of My Adult Life (Extreme Moral Value)

If I conclude that the developmental differences between adult human beings and human embryos are so significant that human embryos are not members of the moral community, then I risk the extreme moral disvalue that my own adult life is precluded. If I were sure that the qualities were morally relevant, then there would be no risk of extreme moral disvalue. If I conclude that the developmental differences are not so significant, then I risk no more than that my adult life is not precluded when it may have been so. But if I cannot conclude that those developmental differences are relevant in my own case, then I cannot conclude that they are relevant in any other case. We arrive at the conclusion that the developmental differences are not relevant in any case. This is enough to establish the greater inclusiveness thesis.

[9]Compare Holland, S., Lebacqz, K., and Zoloth, L., eds. (2001) *The Human Embryonic Stem Cell Debate*. MIT Press, Cambridge. Among the extraordinary potential of stem cells is allowing permanent repair of failing organs. Medical therapy could not only halt the progression of chronic disease but also restore entirely lost organ function. Patients with spinal cord injury, for instance, could receive cell-based treatments restoring central nervous system functions.

Abstract

In the field of regenerative medicine, embryonic stem cell research holds far-reaching promise in alleviating and preventing an array of debilitating diseases and conditions. Yet the biggest ethical stumbling block continues to be conflicting beliefs about the moral status of the human embryo. This has in turn produced heated debate concerning what constitutes the justifiable sourcing of stem cells, leading to the distinction between use and derivation. This essay addresses a wide range of these conflicting perspectives on the moral status of the embryo and asserts that resolving the controversy is notably hampered due to harmful fixations on aspects of the dispute. The study highlights four distinct fixations. These fixations consist of the following: adhering to an inflexible notion of moral status; seeking a single developmental determinant that warrants moral status either during or after gestation; exaggerating the significance of the embryo at the expense of other morally relevant considerations; and deliberating the issue solely within the contours of Western philosophical principles and values. Although there will most likely be no consensus on the embryo's moral status, discussion can only generate possible dialogue if we free ourselves from these fixations. Doing so would require at minimum cultivating a flexible notion of moral status as well as multiple standards for moral status eligibility. In the final analysis, the real practical concern is one of policy. This concern must carefully consider what constitutes responsible use of excess embryos, embryos that, in the author's opinion, lack sufficient moral status that would entitle them to full protection from manipulation.

Fixations on the Moral Status of the Embryo

Michael Brannigan

Introduction

The engaging irony in embryonic stem cell research is that, in the face of abundant promises in the pioneering field of regenerative medicine, this research constitutes the field's biggest hurdle. It is regenerative medicine's moral Achilles heel, reminiscent of the all-too-familiar discourse surrounding abortion and reproductive technologies such as in vitro fertilization (IVF). Namely, does the human embryo possess moral status? Should it be afforded the same protection as all other entities with moral status, the most fundamental protection being the right to exist? Technically, we are talking about the human blastocyst, at the pre-embryo or early embryo stage in which the ovum, now already fertilized, has divided into cells that are still undifferentiated.[1] Should we include this blastocyst within the moral community? This was a major concern when, after researchers at two renowned universities publicly announced on November 11, 1998, that they created immortalized stem cell lines, it became clear to the scientific community that sourcing these cell lines

From: *Biomedical Ethics Reviews: Stem Cell Research*
Edited by: J. M. Humber and R. F. Almeder © Humana Press Inc., Totowa, NJ

would be inescapably controversial. John Gearhart, at Johns Hopkins, acquired these stem cell lines from the tissues of already-aborted fetuses. The University of Wisconsin's James Thompson obtained these stem cell lines from spare embryos as a result of IVF.[2]

What is the morally legitimate source and method for obtaining these pluripotent stem cells, the "master cells" that can later develop into more specialized stem cells and can thereby regenerate into all the various tissue types in the human body? This question has led to what many think of as a crucial distinction between "use" and "derivation." That is, "using" stem cells from embryos that, having already been created, are then ready to be disposed of as in the case with IVF procedures and aborted fetuses, is viewed as morally distinct from "deriving" stem cells by creating embryos for the purpose of generating these pluripotent stem cells. Whether or not this use/derivation distinction is philosophically valid and morally legitimate has been a hot topic of debate and indelibly raises the specter of the moral status of these same cells.

We are familiar with the extreme postures. For many, the human blastocyst is worthy of moral status on the basis of its either being already a human entity or that it has the potential for becoming a human being. In which case, no degree of real medical benefits would justify tampering with the blastocyst. For many others, this clump of human cells is simply that and can be justifiably manipulated or disposed of for whatever reasons. I do not think that responsible scientists share this latter position. All whom I have met treat the blastocyst with respect because it is still a life form and a precursor to a human being. In my opinion, this entire issue has evoked a number of conspicuously unhealthy fixations. Psychoanalysis reminds us of how certain fixations inspire neuroses, whether it is Freud's fixation on the oral stage of development or Lacan's fixation on the premirror stage. Fixations have a way of fragmenting reality on account of the myopic inability to see the broader, all-encompassing picture, the "forest for the trees." Fixations can thus be intellectual in that they adversely affect our levels of understanding. I believe this to be the case concerning perspectives on the moral status of the embryo. These fixations hinder the genuine conversation and dialogue that is necessary to resolve the embryonic stem cell controversy.

Moral Status Fixation

In addressing the moral status of the embryo, we can first become fixed on the notion of moral status in a way that makes this notion itself static. Through indicating moral standing, a moral posture within the community of all other entities possessing moral status, we tend to think of moral status as simply being synonymous with the ownership of moral rights. This fixation enables the straightforward equation: moral status equals having moral rights.

Now, there is no more fundamental moral right than the right to exist. In asserting this, I am not saying that biological existence is what is most important or intrinsically of highest value. I am stating that the moral right to exist is fundamental because any other moral right that we may have rests first upon the right to exist to exercise that right. The moral right to exist is therefore a foundational right.

This then extends the equation: moral status equals possessing moral rights equals possessing the right to exist. Fixation occurs when we reify the idea of "moral status" in such a way that its possession automatically and irrefutably entails the ownership of the right to live. Because of this fixation, once we establish that an entity possesses moral status, we necessarily conclude that this same entity has an absolute right to an existence. Period. This is why so much energy is expended to ascertain whether or not an embryo has moral status. Fixation means that whatever we manage to conclude, the issue is settled. Either the embryo has or does not have the right to live.

But this presumes a static character of moral status. Why would we assume such? Why can we not also think of moral status as a process, admitting of degrees of moral standing? This sliding scale view of moral status is not new and has already been suggested by philosophers such as Mary Anne Warren.[3]

Magic Moment Fixation

A recurring fixation rests upon the unfounded assumption that there is simply one single criterion, one necessary and sufficient condition, that entitles an entity to moral status. This is most evident in the abortion dispute as philosophers isolate specific events in fetal development as decisive criteria; establishing this "magic moment" automatically assigns moral status to the fetus. Theologians often do likewise in asserting the significance of ensoulment, or having a soul, as the defining moment.[4] Setting aside this theological milestone, let us apply to the embryo other turning points that have been advanced: life, human life, potentiality, brain activity, sentience, viability, birth, and awareness of self.

According to the generic criterion of life, all life is sacred and therefore to be protected. As a cluster of living cells, the pre-embryo is undeniably a life form. Some would venture to call it an organism and, as long as it is alive, it has moral status. Thus, according to the rigid notion of moral status, a notion we have already pointed out as flawed, this entitles it to an unobstructed existence. Tampering in any way with the pre-embryo that results in its destruction is wrong.

Conceptually, this sacred quality to life sounds nice. It is hard to take issue with the fundamental sanctity of all of life because life is the basis for all that we experience. Life is inherently sacred in that it is the condition for all other goods. Yet, the practical difficulties with this view are obvious. If life's inherent sanctity entails moral status in the above rigid sense, this makes no sense, because it is a fundamental rule of life that for various

life forms and species to exist, other various life forms and species must die. The position is wholly incongruent with the natural
predator–prey relationship.

Moreover, how far do we extend this notion both conceptually and practically? Which organisms warrant protection over
others? Are we speaking here only of human life? We need to be
clearer about what an organism is in view of the complex spectrum of life from the single-cell amoeba to more physiologically
complex structures. How egalitarian can we be in assuming the
sanctity of all life? If we assign moral status to life forms purely
on the basis that they are biologically alive, then this cannot at all
presuppose a rigid and static version of moral status. As pointed
out earlier, the notion of moral status needs to be extended so that
there are gradients of moral status. This then means that life's
sanctity requires, at the most fundamental level, respecting that
life, even if that life needs to be destroyed. We can respect that
which we at the same time kill. Native Americans taught us this
lesson through their time-honored respect for the animals they
hunted and slayed. The next criterion of human life qualifies the
above sacredness of life position and asserts that biologically
alive human beings assume full moral status. The human blastocyst follows upon the heels of the formation of the single-cell
zygote, the result of the fertilization of the human ovum. This
fertilized ovum contains the genotype, or genetic blueprint, of
the later individual who will pass through the stages of embryo,
fetus, and finally, infant. Therefore, upon fertilization, we have a
complete, compact human genotype. This blueprint is the magic
moment.

This deciding factor is the combination of DNA from sper-
matozoon and ovum that results in the genetic blueprint of the
particular human being. The milestone is the presence of a complete human genome. Yet the thought of my sisters who are identical twins raises a red flag. Identical twins prove that the zygote
can split so that two or more distinct individuals with the same genotype result. Nonetheless, if the blastocyst represents a stage where

individuation is present, is the standard then the formation of a complete genetic blueprint in its primordial stage? If so, to be sure, some would assert that the appearance of the "primitive streak," what would later become the spinal cord, is the demarcation. And this appearance takes place around the 14th day after fertilization. If we take the appearance of the primitive streak as a cautionary marker, then the blastocyst has no moral status, whereas the embryo after 14 days does.

The third milestone is potentiality. Many claim that although the embryo is not yet fully a human being, it has the potential to become one. It is this potential that matters. In this case potentiality is equivalent to actuality, potential equals actual, and therefore a human embryo should be accorded the same full moral status as that of an actual human individual.

Consider the logical repercussions of this line of reasoning. Admittedly, the pre-embryo has the potential to develop into an individual human being. But how do we more precisely define potential? How do we distinguish among the "ability," the "capacity," and the "potential" to develop into an individual? Until we clarify this, we are forced to ask: Would not a zygote have the potential? What about an unfertilized ovum? Of course, the unfertilized ovum does not have a complete human genome that could later develop into an individual, yet the unfertilized ovum does have the potential to become a fertilized ovum. This chain of reduction ends up begging the question. Even if we concede that a pre-embryo is a potential human being, the fact of the matter is that other parties to the debate are actual human beings: the woman carrying the embryo, the humans whose cells are used in generating the embryo, and the humans who suffer from various debilitating diseases and who stand to gain from research on the embryo. Furthermore, within the broader social context of prevailing cloning research, almost all somatic cells have the "potential" to develop into full human beings.

With the standard of brain activity we have a likely candidate for moral status on the grounds of its consistency with a brain-death criterion that establishes legal and clinical death, a standard particularly useful in the euthanasia controversy. If complete and irreversible dysfunction of the brain stem and cortex, known as decerebrate death, establishes clinical and legal death, then why wouldn't brain activity signal the beginning of human life? Yet neurologists situate the beginning of brain activity at around 6 weeks in gestation. This means that the blastocyst, or early embryo, has no brain function because there is no developing brain. If brain activity is the determinant of moral status, then the pre-embryo lacks moral status.

Another proposed criterion has to do with sentience. Sentience consists of the capacity to feel pleasure and pain. Sentience can only occur with the formation of sensory organs and portions of the nervous system. Most neurophysiologists situate this at around 13 weeks into gestation. Thus, the blastocyst or embryo does not feel either pleasure or pain. This may fly in the face of anti-abortionists who resort to films depicting the writhing, first-semester fetus. Yet philosophers, such as Bonnie Steinbock, point out that these are reflex actions. Sensations are not equivalent to feelings.[5] As for the embryo, it has not yet developed any part of its nervous system to feel pain.

Quite a few commentators point to viability, the capacity of the fetus to exist independently of the mother, as the developmental milestone entitling the fetus to moral status and thus full protection. This capacity to exist apart from the mother may or may not necessitate medical intervention. Because viability occurs around the 24th to the 28th week, this standard pretty well sets the benchmark for the 1973 *Roe v. Wade* decision that afforded third-trimester protection to the fetus. Certainly this criterion makes sense in a culture that stresses independence in various forms. We associate individuality with degrees of independence rather than genetic makeup. In any case, the pre-embryo obviously lacks viability. Thus, according to this criterion, the early embryo, and certainly the blastocyst, are not at all eligible for moral status.

Birth is the legally accepted standard of a human person. Once born, the human being obtains legal rights along with all other already-born human persons. For many, birth also determines when a human being becomes a human person in the moral sense. In this case, birth entitles one to the same moral status as all other human persons. Moreover, birth is a significant social, as well as moral, event. At birth, the child visibly enters the human community and becomes unintentionally enmeshed in a web of human relationships. For many, this relational aspect to birth endows it with moral significance. Needless to say, this criterion excludes the blastocyst and embryo from the community of moral concern.

There are those who situate moral personhood, the attaining of moral status, at a time after birth, when the human being eventually acquires self-awareness. The human being gains moral status when he or she is conscious of himself or herself as a subject. "I have a sense of myself." "This is happening to me." This criterion posits a mental capacity as the ground for moral status, a major step beyond the brain activity criterion posited above.

The moral implications of this are far-reaching if we rely upon the static and flawed notion of moral status. Lacking fundamental self-awareness, the human being does not possess the foundational right to exist. Infanticide then becomes a volatile issue. In any case, this standard certainly precludes an organism, such as the blastocyst or early embryo, in which there is no brain activity whatsoever, let alone self-awareness.

To summarize, the major impediment with these efforts to ascertain an entity's moral status is that they end up unfairly simplifying the rather complex process of human gestation and development. It is altogether naïve to look for a single determining standard for moral status. Discovering the "magic moral moment" makes little sense when we consider that human biological development is a gradual process. There are no abrupt changes from, say, the embryonic phase to the fetal phase, from prebrain activity to brain activity, from presentience to sentience, from previability to viability, and from the 9-month fetus to the newborn baby. If biological development is a measured, gradual process, why would we assume that the attainment of moral personhood, that is, the acquiring of moral status, is, in contrast, both abrupt and immediate? It makes perfect sense to think of the acquisition of moral status as a dynamic process and not as a static event. As I see it, the philosophical notion of personhood is congruent with the biological fact of continuous, progressive process.

Embryo Fixation

The burning question continues to be about the moral status of the human embryo. But why this fixation on the embryo? What about other equally if not more relevant concerns? Suppose we all know for a fact that the human embryo is a human person and that it thereby possesses moral status. This would still not settle the issue. To begin with, if we adopt a more flexible, sliding scale of moral status, this would mean that any assumption of the embryo's moral status does not necessarily imply its right to exist. Second, whatever degree of moral status the pre-embryo has may be overridden by other involved parties, parties who also possess moral status. Because these parties are already born and have developed far beyond the embryo, a sliding scale of moral status would mean that whatever moral rights they have are of a wider scope than those of the embryo.

Who are these other parties? They include the biological parents, particularly the mother. In view of the mother's moral status, she may justifiably provide her informed consent to allow parts of her body, including tissues, to be used for research. In the case of IVF, this includes excess embryos. This also includes donating fetal tissue from an aborted fetus. Other parties are those who stand to gain directly from her informed consent and the ensuing research. These are parties who suffer from debilitating conditions, conditions that stand a reasonable chance of being alleviated through embryonic stem cell research. Because the benefits of embryonic stem cell research clearly outweigh the burdens, the moral status of these other parties clearly has priority over the moral status of the early embryo.

Western Cultural Fixation

A good portion of the literature and debate remains fixed in that it occurs solely within the parameters of Western ethical systems and viewpoints. There is a danger in becoming so fixed on our own Western approach, principles, and values that we ignore more global perspectives. Why must we assume that the principles and values that we uphold are universal? Indeed, studying the perspectives of other cultures forces us to reassess our own views.

To illustrate, Hindus ardently believe that the embryo and fetus, throughout gestation, possess moral status and a strict right to live. However, the reasoning behind this has little to do with determining moments of moral status and consequent inclusion within the moral community. It has to do with the belief that the fetus, at conception, is a reincarnated being, and that this being possesses a karmic inheritance from its past existence. It stands to reason that, given this rationale, in the same way that abortion is unacceptable, manipulating the embryo and destroying it is also unjustified. Granted, this reasoning hinges upon religious beliefs. Yet, the facile divide between religion and philosophy that is so apparent in the West is less so in Asian traditions. Therefore, this also compels us to reassess our own views regarding the nature of the relationship between religious and philosophical reasoning with respect to the early embryo and fetus.

The same can be said for Islamic teachings. Muslims believe that life begins at conception. This in itself commands full protection for the fetus at any stage during gestation. Destroying the embryo is wrong just as abortion is wrong. The only exception is to save the life of the mother, because the mother is the "root" and the fetus is the "offshoot," and at all costs the "root" must be maintained.[6]

Chinese beliefs, however, have less bearing on the status of the early embryo and fetus. Grounded in the more pragmatic design of Confucianism, with its emphasis upon filial piety and family duties, Chinese beliefs are more concerned with feasible ways to limit family size and curb China's exceedingly high population growth. Consistent with Confucianism, the wellbeing of the society takes precedence over individual interests. In contrast with Western cultures, Confucian teachings do not emphasize notions of individualism and individual rights, particularly individuals' reproductive rights. Instead, they underscore duties that promote the good of the family, the group, and society.

Japanese beliefs are syncretic, embodying indigenous Shinto, imported Buddhism, and imported Confucianism. Buddhist teachings regarding the nature and origination of the individual are quite complex. Suffice it to say that most Buddhist texts seem to suggest that the human individual comes into being soon after fertilization. Once there is an individual, moral status accrues. Therefore, destroying the embryo is generally considered wrong. Yet, abortion is common practice in Japan. There are two reasons for this. First, Japanese women have very little access to low-dose contraception pills. Second, high population growth threatens the wellbeing and security of a society constantly tested and strained by its small land size. Thus, memorial services for the *mizuko* ("water children," that is, the aborted fetus) are quite popular in Japan. Buddhist temples offer these services (called *mizuko kuyo*) and help to assuage the shame of parents of aborted fetuses.[7]

Conclusion

We desperately need to free ourselves from the above four fixations in order to acquire a more sound notion of the pre-embryo's moral status. Flexibility, with regard to both the notion of and multiple criteria for moral status, must replace the single standard. However, even if we could free ourselves from these fixations and adopt flexibility, in my estimation the moral status of the embryo still remains an irresolvable issue because it rests upon personal, philosophical, and religious assumptions, conceptions, and beliefs. What then do we do? We still need to establish a sound public policy resolution.

Consider President Bush's August 9, 2001 landmark address on the issue of stem cell research. Although his speech evidently sought to appease many parties, by permitting federal funding on the sole condition that the cell lines were created before August 9, 2001, it ended up satisfying no one in particular. This amounted to about 60 stem cell lines taken from around 64 previously destroyed embryos. Scientists have pointed out the naiveté in thinking that these cell lines were sufficient for both sustained research and differentiation into each one of the numerous cell types. His statement was essentially a political compromise, underscoring on the one hand his pledge to protect the innocent future lives of embryos, while at the same time acknowledging the medical breakthroughs that embryonic stem cell research can offer.

The Bush administration now intends to appoint a new committee to ensure the protection of humans used as research subjects. The committee will be mandated to look into whether or not human embryos should be categorized as human subjects. Not only does this reflect the administration's attempt to ban all research that uses cloning techniques, particularly somatic cell nuclear transfer, but it would also outlaw any such type of human embryo research, even if therapeutic. Moreover, it would fully protect human embryos from inappropriate manipulation, thus assigning them moral status. Because this ban does not apply to the private sector, the private sector has neither federal oversight nor rules as to how the research is to be conducted. As of now, we have sparse data on research currently being conducted by private companies. For all we know, the first human clone may have already been created, though we will only hear about the first "successful" human clone.

Nevertheless, if the category of human subjects is to include human embryos, then other countries will no doubt far outpace the United States in embryological science. The aftershocks of incorporating human embryos into the area of human subjects would lead to increasing oversight of new and promising research in human biology and development, along with a perilous decline of studies in the promising area of reproductive medicine.

Rolf Mengele, the son of the Nazi doctor Josef Mengele, claimed that his father expressed no reservations about experimenting upon the inmates at Auschwitz because "they were going to die anyway."[8] Can we make a similar claim for those "surplus" embryos whose destinies are already foreclosed because they will most likely be disposed of? Herein lies the real pragmatic issue: What is the most responsible use of these excess embryos? How do we fairly honor the dead?

Of course we can sidestep the moral minefield and resort to stem cell research that uses only adult stem cells. For example, researchers in Pittsburgh have shown that adult stem cells are a fertile source for regenerating various tissues. However, determining the effectiveness of adult stem cells still requires parallel, comparative studies with embryonic stem cells. In addition, adult stem cell applications remain limited, whereas research involving embryonic stem cells can potentially regenerate every kind of human body tissue and thus alleviate a wider spectrum of crippling conditions.

As I see it, because excess embryos used in IVF are already destined at some future time for disposal (and there is no clearcut policy regarding this future time in the United States), sound and respectful research on these excess embryos constitutes their responsible use. This is particularly the case because embryonic stem cell research shows far-reaching promise in both alleviating and preventing acute and chronic suffering for millions of human persons. By not taking this step in a cautious, sensitive, and respectful way, we relinquish our compassion, a moral duty that necessitates acting in morally responsible ways to mitigate suffering.

There is thus no way to avoid the issue regarding the moral status of the embryo, to do so is morally irresponsible. Yet, we need to address the question in ways that avoid fixation. Fixation polarizes the issue through a naïve reductionism, making a highly complex matter appear simple. Fixation forecloses discussion and dialogue and, especially because of my contention that final resolution and consensus about the embryo's moral status will not occur, dialogue among all the disciplines is all the more necessary.

Notes and References

[1]Some commentators prefer the term "preimplantation embryo." *See* Green, R. M. (2001) *The Human Embryo Research Debates: Bioethics in the Vortex of Controversy.* Oxford University Press, New York, p. 7.

[2]Thomson, J. A., et al. (1998) Embryonic stem cell lines derived from human blastocysts. *Science* **282**, 1145–1147.

[3]*See* the discussion in Warren, M. A. (1987) *Moral Status: Obligations to Persons and Other Living Things.* Oxford University Press, New York, pp. 87–88.

[4]Despite the apparent conceptual difficulties in maintaining that the standard of moral status is having a soul, major world religions share this view. And even though they disagree among themselves as to when this ensoulment takes place, they are much of one voice in recognizing that ensoulment occurs before birth. Here, matters of faith come into tension with matters of reason. Faith is a vital force in our lives. Philosophically, however, there is little empirical evidence for the notion of ensoulment that does not beg the question.

[5]*See* Steinbock, B. (1992) *Life Before Birth: The Moral and Legal Status of Embryos and Fetuses.* Oxford University Press, New York, p. 58.

[6]The abortion of female fetuses is especially heinous in Muslim society because it represents blatant discrimination against women.

[7]Because many Buddhist temples charge exorbitant fees for these services, this has become a source of controversy. *See* Hardacre, H. (1997) *Marketing the Menacing Fetus in Japan.* University of California Press, Berkeley, CA.

[8]Drum, P. (2002) The morality of embryonic stem cell research. *Sensabilities* **6(1)**, 16–17.

Abstract

In the summer of 2001, President George W. Bush announced his policy on stem cell research. The president approved the use of 72 previously created stem cell lines for federally funded scientific research while prohibiting the creation of new cell lines from viable embryos. One of the moral issues that the policy raises is a question about an apparent inconsistency in the President's position. If it is morally wrong to create new cell lines because of the future unethical destruction of viable embryos, then is it not also wrong to use cell lines that were created from the past unethical destruction of viable embryos? Some might argue that if the creation of a cell line is morally wrong, then using the fruits of the wrong action are morally prohibited as well. I contend that using the results of a wrong action or set of actions is not necessarily forbidden, as has been shown in the early to mid-1990s debate on the use of data from Nazi experiments during World War II. According to many leading Judaic ethicists and scholars, if the living require the data from unethical experiments to live or to have good lives, then it is morally permissible to use them for those purposes. On the same grounds, even if killing embryos for scientific reasons is morally wrong, if they are already dead, then it is morally permissible to use the cell lines or test data that result from the killing.

Nazi Experiments
and Stem Cell Research

Dennis R. Cooley

Introduction

On August 9, 2001, after intensive consultation with experts in the field, US President George W. Bush announced his policy on stem cell research, which allowed for 72 previously created stem cell lines to be used in federally funded scientific research while prohibiting the creation of new cell lines.[1,2] The approved lines were developed from destroyed embryos that could have been implanted formerly into women but were not required for that purpose.

The president's policy is the result of two of his apparent beliefs about morality. First, he appears to believe that generating new cell lines is morally prohibited because of the required killing of viable embryos, which might have become persons if they had been implanted. Moreover, the claim that research is morally permissible on the already developed cell lines seems to stem from the president's second belief that although killing embryos for scientific reasons is morally wrong, if they are already dead,

From: *Biomedical Ethics Reviews: Stem Cell Research*
Edited by: J. M. Humber and R. F. Almeder © Humana Press Inc., Totowa, NJ

then it is permissible to use the cell lines or test data that result from the killing[3,4]

Although the former belief has received considerable ethical scrutiny,[5] the second has been left relatively unexamined.[6] I contend that in considering the moral legitimacy of the second of President Bush's beliefs, the early to mid-1990s debate on the use of data from Nazi experiments offers valuable insight into the issue of using stem cell lines developed from embryos.[7] Many arguments found in that canon are valuable in present day reasoning about the morality of using data or material from unethical sources.[8]

In this work, I examine the strongest possible arguments against using the cell lines that have already been developed. The first relies on the Kantian imperative to respect all persons, whereas the second is a consequentialist argument for deterring utilitarian scientists from performing additional unethical experiments in the future. I will argue that both of these arguments fail, and adopt the position of many leading Judaic ethicists and scholars, that is, if the living require the data from unethical experiments to live or to have good lives, then it is morally permissible to use it for those purposes.[9]

Stipulations

The strongest arguments against the use of the 72 stem cell lines assume that some uncontroversial and some extremely controversial claims are true. For the sake of examining the strength of the following arguments, I will make all the claims axioms. First, I will stipulate that human life begins at conception. This claim is noncontroversially true if all that is meant by it is that after the fertilization of the egg has been completed, there exists a living entity with a full compliment of human DNA, which can replicate into more human cells. However, the opponents of using embryos in research generally imply a much higher moral status for embryos than that of human cells capable of replicating in a certain way. Usually, they elevate the fertilized egg to the status of a full human person.

Although there is absolutely no compelling reason to equate a person—that is, an entity that is a moral agent with all that entails—with a human being as defined above, in the interest of fully examining the issue and presenting the best argument against stem cell research I will also concede this point. For the purposes of this argument, the claim that embryos are morally equivalent to full persons is true.[10]

The third stipulation uncontroversially states that some of the clearest cases of unethical research on human beings were the Nazi experiments of World War II. Although the experiments were intended to produce useful scientific data, the manner in which the trials were performed used the human subjects impermissibly.[11] First, researchers did not obtain informed consent from the human subjects, and it is likely that the researchers never realized the need for such consent. After all, the researchers considered the subjects to be prisoners of war or not fully human, thus negating the need for consent.

Furthermore, even if informed consent had been obtained, the horrible and unnecessary pain and suffering of the research subjects would have made the experiments unethical. The subjects received no advantage at all, although the experiments cost them a great deal of suffering, as we can see from some of the examples of Nazi experiments. In one such case, subjects were covered with blankets, dowsed with water, and then left outside on freezing nights to determine how fast hypothermia would set in.[11]

These trials were conducted because researchers were trying to discover methods to delay and treat hypothermia for military personnel who crashed in the ocean or were immersed as a result of their ship being sunk. Naval personnel were very expensive to train and lose, especially the airmen. Using human subjects in the camps, therefore, was viewed by the Nazi researchers as a cost-effective way of saving more valuable entities.[12] However, even if the trials had resulted in benefits to others, the costs to the subjects would have vastly outweighed the benefits, thereby rendering them unethical on these grounds alone.

Because there is a well-developed body of arguments from the early to mid-1990s against the use of data from the unethical research conducted by the Nazis, I will adapt the arguments to the stem cell research debate. Many, such as Stephen Post and Aaron Ridley, argue that results from "evil" experiments ought not to be used because of the taint from the source of the data and the inability to respect the victims while using the data. They claim that use of data from such sources encourages utilitarian researchers to perform more unethical experiments. If those who argue against using the former data are correct, then it follows that it would be wrong to use the 72 stem cell lines or data from experiments that use them.

However, if it can be proved that it is ethical to use data from the Nazi experiments, it then would seem to follow that it must be ethical to use the 72 stem cell lines. I have stipulated that the embryos have the same moral status, that of persons, as the human subjects that suffered through the Nazi experiments. Hence, what is ethical for one group must be ethical for the other, *ceterus paribus*, or with all other factors unchanged.

Moreover, when the moral status of the embryos destroyed by developing the cell lines is juxtaposed against that of the human subjects used in the Nazi experiments, it is clear to most people that the embryos have a lower moral status than that of the human subjects. There is no controversy over the latter's status, whereas the former's is a matter of some debate. If we can prove that it is morally acceptable to use the data from the exploitative and harmful experiments performed on human persons then there will be overwhelming evidence for the permissibility of using data from human subjects that were less than full persons.

Although others, such as the Catholic Church, have resorted to religious texts to bolster their claims, I will not use any form of Divine Command Theory in the arguments because of its many problems. One of the major problems, which renders it inadequate as a source of evidence, is the matter of unresolved, competing interpretation of biblical text. Consider, for instance, several of the interpretations that can be given of the Bible's position on the moral status of fetuses. In Exodus 21:22–25, the loss of a fetus in a physical altercation is to be compensated for monetarily by the offender, which implies that the fetus is the physical property of the father. Bolstering the claim that the fetus is property, the same chapter clearly commands that the unintentional killing of a person should be punished with banishment for the offender (Exodus 21:13).[13] If the fetus has the same moral status as a person from the moment of conception, then the difference in punishments is inconsistent. Hence, the fetus does not have the same moral status as a person does. In the face of abortion opponents, who I assume will reinterpret the cited passages to support their position, a strong biblical case can thereby be made for the permissibility of abortion, as long as the father agrees to the loss of his property.

Regardless of which interpretation is correct, the existence of multiple, contradictory interpretations, for which there are equivalently compelling evidence, should make us reluctant to adopt a Divine Command Theory stance. Fortunately, we do not need to spend a great deal of time and resources, perhaps futilely, to discover the correct reading of Scripture, because we can use other moral theories that are not quite as open to personal bias. Later in this work, I will convert one argument from the divine command position for the use of data from unethical sources into a utilitarian and Kantian argument.

Arguments Against the Use of Data From Unethical Experiments

There are two main arguments against the use of data from unethical experiments: the first is mainly Kantian in nature, whereas the latter is primarily utilitarian. I will develop each in turn, and then argue that it is inadequate to justify the conclusion that we are morally prohibited from using data obtained from unethical experiments.[14,15]

First, some argue that we ought not to use the data from the experiments because doing so is disrespectful to the victims. In fact, for this argument to be valid, we must go so far as to claim that by using the data, we must disrespect the human subjects who suffered so much, otherwise it might be possible for the conclusion to be false while the premises are true. Stephen Post, for example, states that the use of tainted data, or data obtained from evil sources, is "morally revolting" or an "abomination."[15] These regrettably histrionic classifications show that some people believe that we continue to victimize the human subjects merely by using data from tainted experiments instead of leaving them in peace, much like reporters intruding on the suffering of a rape victim so that they can obtain exclusive footage for their news organization and increase their ratings and revenues.

The connection between the use of the data and the disrespect for exploited subjects, however, is not a necessary one as is required for a valid argument. I take "X is disrespectful of Y" to mean that X treats Y as a mere means—not as an autonomous agent—in either thought or deed.[16] The mere fact that the data from an unethical and morally repugnant experiment is used does not necessarily entail that the person, who uses the information, disrespects the victims of the experiments. In fact, he or she could have a great deal of respect for the human subjects who were mistreated with such evil. First, the researcher's behavior does not affect the victims of the experiment, unless the latter allow it to affect them, because the researcher is not using the human subjects in additional trials. The former subjects might think that they are being disrespected, but the belief that one is not being treated as one ought to be is an inadequate indication of wrong action on anyone's part, as long as the researcher's behavior and mental states are above reproach. The victim of a rape, for example, might understandably feel as if the police personnel asking her invasive personal questions are not treating her as an end in herself, even if they are acting in accord with the moral dictates of their profession and moral agency.[17] Though the victims of tainted experiments may continue to feel upset, their subjective feelings do not necessarily entail that anyone has done anything objectively wrong or has rationally validated the feeling. I may make myself upset over not winning the lottery, for instance, but if the result is fair, then there is no objective reason justifying my response.[18]

Furthermore, according to Kantian theory, we are obligated to have the proper feelings of respect towards the victims of the experiments.[16] If the researcher feels the proper emotional response (which will include horror at how the victims were treated) empathy and respect for them as autonomous individuals, and so on, then there is no reason to believe that the victims are being disrespected by use of the data.[19] Hence, there is no necessary connection between the use of data and the treatment of the human subjects of Nazi experiments because it is possible for researchers to have the proper feelings of respect, therefore not affecting the former subjects with disrespectful behavior. In this case the argument based on Kantian theory fails.[20,21]

Aaron Ridley formulates a utilitarian argument against using the data from unethical experiments.[22,23] He believes that "using results from unethical experiments in medicine is wrong because it encourages the performance of further unethical experiments." Encouragement is defined as effectively incited, effectively makes more likely, or effectively promotes further unethical behavior.[22] According to Ridley, encouragement in his argument is a conceptual connection rather than a causal one between the use of the results from past unethical experiments and future unethical experiments being conducted.[22] For the purposes of examining the strongest argument against the use of the 72 stem cell lines, it will be helpful to develop Ridley's utilitarian argument in greater depth.

First, Ridley assumes that there might be scientists who are completely motivated by purely utilitarian considerations. Utilitarian researchers are willing to conduct what most people consider to be unethical experiments on human subjects, which entails that they do not obtain informed consent, appropriately respect the subjects, or act justly, as long as the trial maximizes utility. An experiment that violates other moral principles, with the exception of utilitarianism, is not unethical in the mind of utilitarian scientists because performing the trials has the best utility of all alternatives open to them at that time. In other words, the ends will justify the means, which they will use to achieve the best ends.

Furthermore, utilitarian researchers feel obligated to do the best that they can regardless of how it might affect them personally. In fact, they realize that they might have to bear the burden of severe punishment for their actions, but if the rewards from performing an unethical experiment outweigh the consequences, the burden must be borne. These scientists know that they may even have to sacrifice the interests of their loved ones if doing so is for the good of the whole.[22]

But in the interest of fairness, I must be careful not to suggest that Ridley adopts a purely utilitarian position for his renegade researcher. Ridley is more interested in the motivations of the scientist rather than in the pure consequences of alternatives. His researcher is motivated by the desire to maximize benefit over cost, but that does not entail that the scientist's experiment will actually be the best thing that he or she can do. In order to begin the unethical trial these scientists will first rationally weigh the possible costs and benefits that they are aware of. Their decision will then be made within the framework of utilitarian rules of thumb. Given their evidence and the manner in which events have occurred in the past, they rationally believe that their research will maximize utility. Utilitarian researchers fully realize that these human subjects, and others, may be harmed by what they do. Yet if their experiment produces sufficient benefits over costs, in comparison with the alternatives, then it is an experiment that they are morally required to perform.

One of the conditions necessary for the researchers to make a rational decision on whether or not to conduct the experiment is that there be a social policy or convention in place which states that the data from unethical, but scientifically valid, experiments can be used by the society. On the other hand, if there is a rule not to allow the use of the data, then the researchers, being good utilitarians, will see no value in performing the research. Clearly, if the information will not be used, there will be costs with no benefit. But if data from such experiments are generally utilized, then they will rationally perceive that this data will likely be used as well, thereby justifying their utilitarian decision to proceed with the exploitative research.

According to Ridley, it follows from the policy of using the results of unethical experiments that we encourage utilitarian researchers to perform additional unethical trials. Because the utilitarian calculus shows that certain unethical trials will produce the best overall result, we increase the chances of them being performed, rather than taking the opportunity to guarantee that no more will be conducted. Never using the results would entail that utilitarian-motivated researchers would rationally decide to never perform unethical experiments, for they would never lead to the best overall outcome. Ridley's conclusion is that we are obligated never to use the results from unethical experiments, no matter how tempting such use is.[24]

There are several responses to the argument, of varying force, that I will now consider, starting with the weakest and proceeding to the strongest. First, it might be the case that unethical experimentation can be prevented by methods other than not using the data from morally flawed trials. If it is made abundantly clear that the self-interest of the unethical researcher will be greatly harmed, then the research community might go a long way in preventing unethical experiments while still being able to use the data from Nazi experimentation. J. David Bleich claims that "[p]ublication is probably the primary goal motivating scientists to engage in medical research."[25] If the results of unethical experiments are published at the same time the researchers are censured and their methods repudiated by the editor, then there are strong disincentives to future lapses of ethical conduct in human subject research. Furthermore, "[s]ince enhancement of scientific reputation as well as ego gratification are a function of frequent citation in scientific and medical literature, it should become the accepted norm for ethically tainted studies to be cited without attribution of authorship." Hence, it is better for researchers to be ethical rather than unethical so that their individual interests are best served.

The problem with adopting Bleich's position as a response to Ridley's argument is that there are different theories at play, even though they both focus on consequentialism. Ridley is using a utilitarian theory, whereas Bleich has adopted egoism. If scientists are concerned for their own benefit, but are more committed to utilitarian ideals, then they will not be swayed merely by considerations of self-interest. They will do what they perceive to be the best for everyone, even if that will lead to the worst outcome for them. Therefore, though the sanctions that Bleich recommends will prove effective in a great number of cases, they will merely be minor considerations in the calculus of a devout utilitarian researcher.

A stronger response to Ridley's argument is to claim that he does not understand how experiments are performed in the real world. Generally, experiments do not produce results that would justify the unethical use of human subjects, which is something that any adequate researcher would already know. The results of most experiments are insufficiently life altering for a sufficient number of people to justify going outside the bounds of moral propriety because scientists must study small parts of a very large problem or first determine what does not work rather than what is efficacious. The large majority of research is not involved with testing a cure for one of the major diseases, such as cancer, AIDS, and so on, that would justify violating ethical guidelines of conduct.

Research takes a great deal of time and resources to produce even minor advances. It is therefore not wise, on utilitarian grounds, to perform unethical trials, because doing so will eliminate the ability of the researcher to perform further research. Utilitarian scientists are faced with the likelihood that either the resources that allow them to perform experiments will evaporate because of increased public scrutiny and censure, or they will be sent to jail, where it will be much more difficult for them to conduct their research. Of further consideration in the cost–benefit analysis is that if scientists are prevented from carrying out their work early on because of their experimental conduct, then the loss of knowledgeable personnel would be costly for research in this area overall until adequate replacements could be found. Hence, a rational utilitarian researcher will know about these limitations and refrain from conducting research that will not lead to the best overall results.

Ridley might respond that some research is significant enough that it would justify the use of human subjects even if such use were unethical on other moral grounds. To begin research in certain areas, or to have any research at all, the only method might be to perform unethical experiments. If we, for example, wish to truly discover how the human brain develops or functions, then it might be necessary to implant measuring devices, that are presently used only in animal brains, in humans. It is informative, for example, to see what a monkey's brain will do when stimulated in certain ways, yet seeing information from actual human brain stimulation would yield more accurate information because monkeys are significantly different from human beings. The data from the experiment might be necessary to tell us what we need to know and, though unethical, may be the only way in which the evidence could be found. This consideration could be an inducement significant enough to make the utilitarian researcher decide to perform the unethical experiment.

An effective response to Ridley's intriguing argument is the problem of mediated consequences that all consequentialist theories encounter.

Roughly, because the moral value of an action rests solely upon its consequences, the action's value can be changed, from right to wrong or wrong to right, by others responding to or mediating the original action's consequences. A seemingly right action might in fact turn out to be the worst thing one can do, and vice versa because of how others mediate the consequences of the action. For example, suppose that Hitler fell into a stream when he was a child, and someone pulled him out. Furthermore, suppose that he would not have been rescued any other way and, of course, he does not die until he commits suicide in his Berlin bunker in 1945. On utilitarian grounds saving Hitler's life as a child would be morally wrong because of what he did as an adult. It would have been better on purely utilitarian grounds if he had died before he committed his crimes against humanity. Hence, though the rescuer's action appeared moral at the time, it was unethical because of what others did as a result of Hitler's life being spared.

When Ridley claims that we encourage others to commit further unethical trials by using the results of unethical experiments, he opens the door to this problem. Even though his theory does not use the actual consequences of the alternatives in the utilitarian calculation, it does make the morality of an action entirely dependent upon what the agent performing the action perceives to be the likely consequences of the alternatives. Hence, if there is good reason to believe that someone will mediate the consequences of another's seemingly right action so that it does not maximize utility, then the agent is prohibited from performing that action. The same idea applies to seemingly wrong actions that are mediated by others to maximize utility. In consequence, an agent's actions can be manipulated by what others do, regardless of the agent's actual desires and intentions.

However, it is clear that the morality of our actions is not always dependent upon what others do as a result of our actions, even if we know with almost absolute certainty what will occur. Suppose terrorists kidnap innocent children, kill one to show that they are willing to kill the others, and then inform us that they will kill the others one at a time if a large ransom is not paid immediately. Because we do now wish to participate in the evil scheme for whatever moral reasons we might have, we do not pay the ransom. Our action conceptually encourages the terrorists to kill all the children, which they promptly do.[24] According to Ridley's version of utilitarianism, our refusal to pay is unethical because it did not lead to the best outcome.

However, the mere fact that deaths occurred as a result of our action, even if it is a mere conceptual connection, does not entail that our refusal to pay the ransom is morally wrong nor does it make us responsible for the murders. The wrong actions and blame for the deaths clearly lie solely at the feet of the murderers, who illicitly mediated the consequences of our action. Their actions are wrong, while ours are permissible.

On the same grounds, the scientific community can act ethically by utilizing data from unethical trials even though it encourages researchers who violate ethical conduct principles to perform additional unethical experiments. Although their actions are a foreseen consequence of the data use, their distinctness from the community's action proves that they have no bearing on the moral status of the community's action, unless the unethical trials are a result of the community's intentional actions or negligence.

Ridley's error seems to be based on a lack of awareness of two types of encouragement: one in which moral responsibility for others' mediating actions can be assigned to the person doing the encouraging and a second in which no responsibility is attached. Each is listed below.

Intentional or negligent encouragement = df. We know
or have good reason to believe that X will result as a mediated consequence of our actions; X is wrong or bad, and we
desire or do not care if X results. The third condition makes
us morally responsible for this type of encouragement and
the mediating actions that we intentionally or negligently
encouraged to result.

Unintentional encouragement = df. We know or have
good reason to believe that X will result as a mediated consequence of our actions, X is wrong or bad, and we truly
want X not to result. No moral responsibility is attached to
this type of encouragement or the mediating actions that
result.

It is beneficial to see how these two types of encouragement
work and why responsibility is attached in one but not the other.
Suppose that a racist, who has encountered very few blacks in his
lifetime, sees a black man with a white woman, who is obviously
the man's significant other. The encounter with the couple serves
to reinforce the racist's stereotype of black men's desire for white
women, which causes him to discriminate against blacks in the
future. Under Ridley's theory, the black man has done the wrong
thing because he has encouraged the racist to strengthen his unjustified stereotype and act as a racist. However, the encouragement
is unintentional, so the man is not morally responsible for what
the racist thinks or does.

However, if the black man was going out with a white
woman primarily to reinforce the stereotype or does not care if
black men will continue to be stereotyped in this way, then he has
done something wrong. He, in effect, conspires with the racist to
produce the unethical result and future racist actions.

The encouragement that Ridley describes in his argument
against using data from unethical trials is unintentional rather than
intentional encouragement. The scientific community clearly
does not want others to perform unethical experiments and is will-

ing to do without data obtained from them. The only way that the scientific community would be at fault for encouraging evil research would be if it intentionally or negligently created a policy to use the data which it reasonably believed would produce additional unethical experiments. Such a policy would enable researchers to obtain the data resulting from such experminentation. Because this does not seem to be the case—although it is an empirical question that needs to be answered—Ridley's argument fails on these grounds alone.

In addition to inadequately defining encouragement, Ridley fails to grasp the fact that people who perform unethical experiments may not make decisions rationally. There is no reason to believe that rational decision procedures, intended to inform them not to do something unless it will maximize utility (according to their adopted consequentalist theory), will be followed. That is, Ridley assumes that people will calculate utility rationally, without bringing unjustified hope into their calculations. However, many people acting unethically seem to think that they can get away with their crime, even when it is virtually certain to most rational agents that the former will be caught.

The same result holds for the utilitarian researchers that Ridley describes. A researcher can acknowledge that there is a policy against publishing or using illicitly obtained data yet still be convinced that the results of his study will be so important that they will cause an exception to or a reversal in the policy. Clearly, the policies of the country and society have changed over time in light of new evidence, so it would be rational for him to believe that he would succeed where others had failed, with publication of his data set. The conclusion will form a part of his calculations, driving him to commit the most unethical of experiments in hope that the results would be so extraordinary that they would be an exception to the absolute rule society had previously adopted toward unethical experiments.

The only way to eliminate this result is to eliminate the hope of the researcher, but that would be difficult at best, if not impossible. It would require crushing rational and irrational hope in the researcher. Moreover, eliminating hope may destroy what many consider to be a necessary condition of human nature, which may be unethical on its own grounds.

A Positive Argument

The mere fact that a data set stems from unethical experiments or sources neither shows that the use of the data is inherently or prima facie wrong, nor that it encourages utilitarian researchers to conduct further unethical experiments because doing so is likely to maximize utility. Moreover, there is a positive argument that can be formulated that proves that the use of data from tainted experiments is morally permissible in certain cases. The argument that I will develop is based upon a position found in the Judaic response to the unethical experiments conducted by the Nazis in World War II.

Barry Freundel, among others, asks the question, "Given the horror perpetrated by the Nazis, should we grant them even one more victim by preventing the use of what in the final analysis are objective scientific facts concerning nature and its workings?"[26] Freundel argues that the *Choleh Befaneinu* requires that a sick person in need take precedence over any "distant considerations—even heartfelt historical national traumas." That is, if some good can come to the living—or even to future generations—by using the data from unethical experiments, then it is a moral imperative to do so. Refraining from using the data merely harms the ones who can benefit, while not helping those who suffered from the experiments in the first place.

Because I wish to avoid competing claims about the correct interpretation of religious texts, I will use merely the two main moral principles that seem to be incorporated in Freundel's argument and put aside the religious arguments. Freundel seems to incorporate a utilitarian and Kantian theory in his position. First, he calculates the utilities of using and not using the data by comparing the outcomes of who will be harmed and who will be helped. Assisting the living is more important than protecting the dead or those who have suffered through unethical experiments; the former cannot be harmed and the latter are not being harmed further unless they hurt themselves by becoming overly upset. Hence, using the data will produce benefit with little direct cost, while not using it will produce no benefit and a greater cost in pain and suffering.

Second, Freundel states that "Judaism sees each life as infinitely valuable." If a person's life is infinitely valuable, then it must follow that the person, whose life it is, is infinitely valuable, *ceterus paribus*. It would be an implausible result if a necessary feature of what it is to be a person has infinite value when the whole does not. The loss of value would have to be explained in some satisfactory way, whereas the maintenance of value for the whole does not. Hence, because people are infinitely valuable, and there is a duty to respect the infinitely valuable, we are obligated to respect people, a Kantian Categorical Imperative to not treat anyone as a mere means.[27]

I will state the versions of utilitarianism and the Categorical Imperative that I have adopted, each of which must be satisfied by an action in order for the overall theory to classify the action as ethical. The proof for the legitimacy of these two subprinciples has been stated elsewhere in more depth.[28] The first is what I call *reasonable person utilitarianism* (RPU). The theory states the following:

> **RPU** = df. An action is morally right only if a reasonable person would reasonably believe that the action would probably maximize utility.[29]

RPU is a practical rule of utilitarian theory. Although following it will generally maximize utility in most cases—and be best overall for everyone—it will not maximize utility in every case. People, after all, make errors in their utility calculations, other agents may mediate the consequences, or some other unforeseen and unforeseeable event could occur that will have an impact on the actual utility of the action. However, the morality of an action depends, in part, on the consequences that the agent can reasonably believe will result; hence, there will be different classifications of alternatives for RPU and any standard act utilitarian theory.

The second theory is a version of Kant's Categorical Imperative:

Quasi-Kantian Categorical Imperative = df. An action is morally right only if in doing the action, the agent does not treat anyone as a mere means.

I take Kant to mean that respecting a person requires not only appropriate behavior, but morally correct mental states as well. The primary motivations, intentions, and mental attitudes to the person should be morally correct, such as virtuous motivations, good intentions, and the mental attitude of respect for all persons affected by the action.

The theories would classify the use of data from Nazi experiments as ethical, although the research itself is unethical. First, using the data is probably going to maximize utility over not using the data. There are many experiments that cannot be repeated because of their unethical nature, so the ones that have legitimate data could give new ideas and information that can never be replicated, given current technology and moral prohibitions. Moreover, because the experiments have been completed, time and resources need not be expended again on finding out what we already know. There is also a huge potential for benefiting a number of people, either now or in the future, as we develop new medical products based, in part, on the data. Finally, the only real cost is how others view the use of the data. If they are upset about the utilization in sufficient quantities, then their negative feelings would alter the calculation of probable uses. However, the distress that they would feel does not seem to be as severe as the pain and suffering that would be caused by not using the data. Hence, utilizing the data probably would produce the best outcome.

Second, if the researchers using the data respect everyone affected by their actions as ends in themselves, then they have satisfied the Quasi-Kantian Categorical Imperative condition. All that they need to do is to exhibit proper behavior and have the correct mental states. As was shown above, the mere use of the data does not entail that anyone is disrespected. The data can be used by a researcher while he or she simultaneously instantiates the proper attitudes toward the methods used in the unethical experiments, such as condemnation of the sources and the unethical researchers, and respect for those who were treated with such evil. Hence, if both conditions of the overall theory are met, the use of data from unethical Nazi experiments is morally permissible. Furthermore, on the grounds of the morally relevant similarities between the Nazi experiments and stem cell research, it follows that the use of stem cell lines is ethical as well, which justifies President Bush's position on stem cell research.

Conclusion

In an ironic way, the Nazi experiments are useful for more than the data that were obtained from them. First, they are paradigm cases with which to teach researchers about ethical research and why one ought not to conduct research in certain ways. Second, the philosophical discussions that they have generated help to provide insight into an area of research that could only have been imagined, at best, in the 1940s. Hence, it does not follow that good cannot result from an action or pattern of behavior that is clearly and indubitably unethical.

Acknowledgment

I would like to thank James Humber for his valuable suggestions for improving this work.

Notes and References

[1]Canada and the United Kingdom have approved the development of new lines of stem cells from human embryos. (*See* Kondro, W. and Holden, C. (2002) Canada gives ok for new cell lines. *Science* **295(5561)**, 1816.)

[2]Most funding of stem cell research in the United States comes from private, rather than public, sources. *(See* Berger, E. [2001] Research Avenue Adds Fuel to Stem Cell Controversy, CNN, July 11, 2001 200wysiwyg://222/http://css.allpolitics.prin...7832467 983019005&partnered=2001expire=-1.)

[3]Other countries take a more liberal approach to stem cell research. Britain is likely to establish a stem cell bank, according to the Medical Research Council's plan, revealed last week, to create a bank containing cells harvested from hundreds of thousands of healthy human embryos, which will thus be destroyed. Politicians are delighted that this will make Britain the world leader in a pioneering branch of medical research, as it is the only country that allows so much access to embryonic stem cells for those seeking to cure diseases ranging from Alzheimer's to Parkinson's. *See* Anonymous. (2002) Go-ahead for stem cell research on embryos. *Telegraph* February 27, 2002, http://news.independent.co.uk/world/ science_medical/story.jsp?story=329216.

[4]There is a possible cell source that might eliminate the need for cells from human embryos. Advance Cell Technology claimed that it had chemically induced parthenogenesis on monkey oocytes, which might also be induced on human oocytes. *See* Holden, C. (2002) Stem Cell research: primate parthenotes yield stem cells. *Science* **295(5556)**, 779–780.

[7]The issue of Nazi experimentation was sufficiently powerful to generate a conference in December 1989 at Boston University attended

by the likes of George J. Annas, Michael A. Grodon, Arthur Caplan, and Elie Wiesel, among other notables (*See* Annas, G. J. and Grodin, M. A. [1990] The Nazi doctors and the Nuremberg Code: relevance for modern medical research," *Medicine and War* **6**, 120–123.)

[8] I will leave aside the problems of too few stem cell lines and difficulty in distributing the 72 cell lines (*See* Vogel, G. [2002] Are any two cell lines the same? *Science* **295[5561]**, 1820.)

[9] Israeli scientists have helped to set the pace for the rest of the world on stem cell research. *See* Vogel, G. (2002) In the Mideast, pushing back the stem cell frontier, *Science* **295(5561)**, 1818.

[10] Some philosophers argue that the embryo cannot have an identity because of a lack of brain cells and habits (McGee and Caplan, p. 154). I find this argument convincing, but put it aside for the purpose of this work.

[11] Robert L. Berger contends that the trials were so incompetently performed from a scientific viewpoint that they yielded no useful data, which eliminates the need to discuss the morality of using such data (*See* Berger, R. L. [1992[Nazi Science: comments on the validation of the Dachau human hypothermia experiments, in *When Medicine Went Mad.* Humana Press, Totowa, NJ, pp. 109–133.) On the other hand, Kristine Moe, Jay Katz and Robert S. Pozos argue that the data are scientifically useful. *(See* Moe, K. (1984) Should the Nazi research data be cited? *The Hastings Center Report* **[December]**, pp. 5–7; Katz, J. and Pozos, R. S. [1992] The Dachau hypothermia study: an ethical and scientific commentary, in *When Medicine Went Mad.* Humana Press, Totowa, pp. 135–139; and Pozos, R. S. [1992] Scientific inquiry and ethics, the Dachau data, in *When Medicine Went Mad.* Humana Press, Totowa, NJ, pp. 95–108.) Hence, a discussion of the morality of using the data is warranted.

[12] The emaciation of the human subjects made the results of the experiment of questionable value considering that the servicemen were healthy human beings. *See* Robert L. Berger's article "Nazi Science."

[13] *The Holy Bible: Revised Standard Edition* (1974) The Penguin Group, New York.

[14]Baruch C. Cohen's "The Ethics of Using Medical Data from Nazi Experiments" provides an excellent overview of the problem as well as listing some of the more horrendous experiments (*See* Cohen, B. C. [2002] The ethics of using medical data From Nazi experiments. *Jewish Law-Articles* [**February**], 1–36.)

[15]Stephen G. Post claims that "[b]ecause the Nazi experiments on human beings were so appallingly unethical, it follows, *prima facie*, that use of their results is unethical." (*See* Post, S. G. [1991] The echo of Nuremberg: Nazi data and ethics. *Journal of Medical Ethics* **17**, 42–44.) However, this conclusion seems to be a mere intuition on Post's part. Many good things result from appallingly unethical events, and just because an action stems from an evil event does not entail that it is tainted. For example, if one person gets a job because the other candidate is murdered, then it follows from Post's account that accepting the job would be prima facie unethical. But it is not at all clear that this is true. Post needs to present an argument that shows the distinction between the two, which is not based on pure intuition.

[16]Kant, I. (1996) *The Metaphysics of Morals*, trans. and ed. by Mary Gregor. Cambridge University Press, Cambridge, pp. 23–24, 198–199.

[17]I assume that acting according to the morality of one's profession simultaneously satisfies more general moral theories which govern all actions.

[18]The lottery example is not intended to minimize the pain and suffering experienced by the victims of unethical experiments. Rather, it is merely intended to make the point that it is irrational to victimize oneself when one does not have to do so.

[19]Mark Weitzman makes the same point when he writes "We must never forget the victims, and we must use the material with the full acknowledgement of the moral flaws in its procurement" (p. 30). *See* Weitzman, M. (1990) The ethics of using Nazi medical data, a Jewish perspective, *Second Opinion* **14**, 26–38.

[20]It is controversial to say this, but if the victims have died, then it is impossible to disrespect them. The most one can do is to disrespect their memory, legacy, or their ordeals, which still exist in some way, while they as persons cannot exist. There is no such thing as a dead person. There is no way to disrespect something that does not exist.

[21]Others might argue that the use of the data is wrong because it condones the unethical methods employed in the experiment. By not using the data, we send the clear message that such research is morally wrong and can never be made right. However, as in the case of disrespecting the victims of unethical experiments, there is no necessary connection here between condoning the unethical methods and using the data either. There are many situations in which moral agents have to react to what others have done. In *Sophie's Choice*, for example, Sophie is forced to choose, by a Nazi concentration camp guard, which of her two children will live. If she does not choose either, then both will be killed. She selects the one she thinks has the best chance to survive in the camp. Her participation in the selection, nevertheless, does not entail that she condones what the guard is doing. She is merely making the best of an overall evil situation.

[22]What Ridley means by the term "unethical experiment" is any experiment that does not satisfy the current guidelines of obtaining informed consent, equity in subject selection, etc. For the purpose of consistency, I will adopt his definition of the term. *See* Ridley, A. (1995) Ill-gotten gains: on the use of results from unethical experiments in medicine, *Public Affairs Quarterly* **9(3)**, 253–266.

[23]Benjamin Freedman presents and rejects a similar argument that more clearly relies on a causal connection between the use of data and the encouragement of future experiments. *See* Freedman, B. (1992) Moral analysis and the use of Nazi experimental results, in *When Medicine Went Mad*. Humana Press, Totowa, NJ, pp. 141–154.

[24]The notion of conceptual encouragement is problematic at best. Encouragement is causal in nature because when one person encourages another he is trying to bring about a certain effect, such as making the person feel better, try harder, etc. Hence, conceptual encouragement seems to be an empty term.

[25]*See* page 73 in Bleich, J. D. (1991) Survey of recent Halakhic periodical literature, utilization of scientific data obtained through immoral experimentation. *TRADITION: A Journal of Orthodox Thought* **26(1)**, 65–79.

[26]Freundel, B. (1995) Using Nazi experimental data—Asking the right

question, *Sh'ma, A Journal of Jewish Responsibility* **25(490)**, 1–3.

[27]The value of human life could also be used in a utilitarian calculation. If lives could be saved by using the data, while they could not if the data are not used, then it follows that using the data is the best thing to do.

[28]Cooley, D. (2000) Good enough for the third world, *The Journal of Medicine and Philosophy* **25(4)**, 427–450

[29]The problem of defining a reasonable person is too large of a task for any essay. I will merely stipulate that a reasonable person is one that uses the moral theories correctly given his information, time constraints, ability to reason, and other situational factors.

Abstract

When we think of the state of bioethics at the present time, the moral dimensions of stem cell research and cloning, genetic therapy and counseling, reproductive interventions, organ and tissue transplantation, and life-prolonging (sometimes better described as dying-prolonging) technologies come to mind. All of these issues have come to the forefront because of the incredible and almost unimaginable advances in biology, chemistry, physics, and electronics over the past quarter of a century, which have greatly increased the potential for enhancing human wellbeing. Our cultural values and laws have barely been able to deal, in a coherent way, with the opportunities and dangers posed by the advances which are resulting from this accelerated pace of scientific development. The media, and people at large, are often supportive of these technologies because, although the creation and destruction of new human life and the transformation of human embryos into stem cells are often the "collateral damage" of this research, the focus is generally on how well off everyone would be if we could just develop the research into new and better therapeutic applications. Scientists working in these fields are almost always asking for government support without any restrictive legal guidelines or prohibitions. I believe that it is up to bioethicists to ask the difficult (and often unpopular) ethical questions that relate to contemporary advances in biomedical research that will guide both scientists and public policy in the years to come.

Recent Ethical Controversies About Stem Cell Research

James J. McCartney

Introduction

In this chapter, I wish to concentrate specifically on the following four topics of ethical controversy relating to stem cell research. First, the allocation of government resources: Given the lack of equitable distribution of health care resources in the United States (not to mention in the world), should any government resources be spent on stem cell research, since its benefits, if any, will only accrue to those who can access provision of health care rather easily and who are able to afford the (at least initially) expensive clinical applications associated with this research? Second, a discussion of the moral status of the embryo with a focus on human rights, as well as my own analysis of this issue based on traditional philosophical and theological teachings. Third, sources of stem cells and the ethical implications of their derivation, and fourth, the relationship of the ethical debate

From: *Biomedical Ethics Reviews: Stem Cell Research*
Edited by: J. M. Humber and R. F. Almeder © Humana Press Inc., Totowa, NJ

surrounding stem cell research to public funding and the development of the law.

By choosing to discuss these aspects of the controversy I will be excluding others, for example, a thorough discussion of the patenting of embryonic stem cell lines, the validity of the informed consent of past embryo donors (whose embryos have been used to develop established cell lines) who believed that their donation was going to be somewhat anonymous,[1] and the issues surrounding therapeutic cloning.[2] I leave it to other authors of this volume to cover some of these topics and others not mentioned here.

Issues of Justice

According to Lucette Lagnado in a poignant article in the *Wall Street Journal* on November 12, 2002, there are 41.2 million Americans who lack health insurance.[3] Many of these are the working poor, but as the number of uninsured continues to increase, it is becoming an increasingly middle class problem. People without health care insurance are left to ration their own health care and act as their own physicians and druggists.[3]

A conference on this issue in April 2002 at Georgetown University Medical School's Center for Clinical Bioethics (2002)[4] described this situation as an "outrage," and participants were encouraged to frame all discussion of bioethical issues within the context of the question: "How does my topic relate to the reality of those with little or no health care insurance?"

This is an especially important "ethical parameter" if we consider the issue of stem cell research. Assuming that most or all of the projected successes of stem cell research do have clinical applications and will be able to save lives, or at least improve the quality of lives ravaged by diseases such as Parkinson's, whose lives will be saved or improved? A societal reshaping of health care distribution (unlikely, given the complete rejection of the Clinton health care plan, which attempted such a reshaping), must take place in order for the clinical applications of stem cell research to be made available to the general public. If this does not occur, then these hoped-for-but-as-yet-unproven benefits will be enjoyed only by those who can afford them, that is the rich and the upper middle classes, who currently enjoy excellent health care coverage complete with prescription drug benefits. This is not only unfortunate but also unfair because it is the tax money of millions of Americans (including the working poor) that will subsidize this research if it is ever approved for full federal funding. Perhaps our tax money might be better spent improving the access to, and quality of, health care for all Americans (almost everyone agrees that this is ethically praiseworthy), instead of allocating it to those performing research on embryonic stem cells; a field of research which is, at best, ethically controversial, and, at worst, in total disregard of the dignity and uniqueness of human life.

I am not suggesting that we abandon basic research that might improve the lot of suffering people (private funding should not be proscribed), but I am recommending a very limited role for federal and state governments in supporting this research with tax dollars. And if the government does make a financial commitment in support of less controversial research, for example, research involving adult stem cells, provision should be made at the outset that the benefits of this research will be distributed more fairly than is the case with most health care benefits at the present time.

If we focus only on the ethics of the techniques for deriving embryonic stem cells and the benefits that successful derivation and clinical research may provide, we will miss the larger picture of the ethics of distribution of health care resources and will once again avoid considering the black cloud of unfair distribution that hangs over all our health care-related activities. In a recent article, Chris Hackler states: "One thing seems certain: rationing of health care will test severely our capacity to subordinate our own private good to the public good."[5] This is especially difficult for researchers, whose reputation and livelihood may depend on their ability to convince the American public and Congress of the importance of their work in the field of embryonic stem cell research (no matter how ethically controversial it is or how it diverts funds that could be used to provide a more equitable distribution of health care resources).

Moral Status of the Embryo

Stem Cells and Cloning

In October 2000, I gave a lecture on the ethics of stem cells to members of The Virginia Biotechnology Association. I was well into the lecture when one of the participants raised the question from the floor, "What exactly is a stem cell?" I realized at that moment that I had violated an important principle that I always try to instill in students: *explain your terms*. It is with that in mind that I will try to explain adult stem cells, embryonic stem cells, cloning, and parthenogenesis insofar as these latter two technologies are or may be used to produce embryonic stem cells.

In the introduction to their book, Holland et al.[6] provide some helpful definitions. Stem cells can be briefly described as unspecialized cells that give rise to various types of specialized cells. In more technical scientific language, they are described as pluripotent cells with the capacity for prolonged self-renewal, which can be induced by chemical and electric means to form different types of specialized somatic cells and tissues (pluripotency is the ability to develop into many kinds of tissues and organs of the body).

Cloning was popularized a few years ago by the production of Dolly, a sheep who died prematurely on February 14, 2003.

A cloned organism is defined more technically as an individual grown from a single somatic cell nucleus of its parent that has been implanted into an egg cell from which the nucleus has been removed; then this enucleated egg cell containing the implanted nucleus taken from the somatic cell is stimulated electrically and cell division and production of a new embryo begins to take place. In a series of articles dealing with stem cell research,[7] cloning is defined as "creating a genetically identical organism, through any of several techniques." Dolly the sheep, the first mammal to be cloned successfully, was created by inserting DNA from the nucleus of a sheep mammary gland cell into an egg cell emptied of its own nuclear DNA.

Somatic cells are any of the cells of the body except sperm cells and egg cells, for example, the udder cell whose nucleus was used to produce Dolly. The clone is genetically identical to the parent whose cell nucleus was used because both share the same genotype (identical nuclear DNA). Parthenogenesis is the stimulation of a not fully mature egg cell (it is still attached to a polar body, which contains the other half of its DNA) to induce cell division and eventual production of stem cells. In one of the previously mentioned articles,[7] parthenogenesis is defined as "[r]eproduction in which the egg develops into an embryo without fertilization. Parthenogenesis does not occur in mammals, but scientists can use chemicals or electricity to stimulate the eggs of certain animals into dividing as if they had been fertilized. One company has started experiments with human eggs."

Advanced Cell Technologies has claimed success in beginning the process both of human cloning and parthenogenesis through a few cell divisions, but it says it intends to use the embryos produced by cloning or parthenogenesis, if any, solely for stem cell research.

Stephen S. Hall points out that most of the cells in the body are specialized to perform their functions within unique tissues and organs, but adult stem cells—found in various locations in the adult body—can function within, and even form, a number of different tissues and organs, and so, in theory, could be used to treat a vast array of diseases.[8] Stem cells are taken from the body by simple removal or extraction. Stem cells that produce blood (and possibly other tissues) can be derived from the umbilical

cord, which is saved after birth and frozen. A Kaiser Daily Reproductive Report has indicated that there is some controversy at the
present time as to who owns these umbilical cord stem cells.[9]

Embryonic stem cells are derived from embryos in the following way: "As the fertilized egg divides, each cell is able to be
separated out and form an entire new organism.... At the point at
which the dividing cells develop into a hollow ball, the embryo is
called a blastocyst. Human embryonic stem cells are derived by
destroying the outer shell of the blastocyst, which would normally
become the placenta, and culturing cells from the inner cell
mass."[6] These cells, which are grown in tissue culture and can
actually develop into organs, are now referred to as pluripotent
rather than totipotent because they can no longer develop into
embryos.

Twinning and Chimera Formation

Two other phenomena should be mentioned at this point:
twinning and chimera formation. Up until the formation of the
primitive streak (what will become the spinal cord) and cell differentiation (about 14 days after fertilization), the developing
embryo can cleave naturally or be cleaved artificially to produce
identical siblings. Hence, embryonic cells that are still part of the
inner cell mass are described as totipotential because they can
give rise to new organisms, that is, twins.[7] Another possibility
during this same timeframe is that two developing embryonic cell
masses with different genotypes can fuse to form what is described
as a chimera. A chimera is an organism whose cells derive from
two or more distinct zygote lineages.[10,11] After the cells have begun
to differentiate and the primitive streak has formed, twinning and
chimera formation are no longer possible. This stage, called individuation, is well described by Norman Ford, a moral philosopher
very knowledgeable about embryology.[12] John F. Kavanaugh
cites relevant materials from Ford's book concerning individuation and observes that "[d]ata concerning this stage are important
in controversies concerning 'morning-after' pills, preimplanta-

tion diagnosis, and 'stem-cell' research."[13] This discussion about individuation will be continued in a later section where it will be integrated into philosophical analyses of human personhood.

Respect for Human Life

Respect for human life is the leitmotif of the *Universal Declaration of Human Rights* (1948) approved more than a half-century ago by the General Assembly of the United Nations.[14] That document states in part that "all human beings are born free and equal in dignity and rights," that "everyone is entitled to all the rights and freedoms set forth in this Declaration, without distinction of any kind, such as race, colour, sex, language, religion, political or other opinion, national or social origin, property, birth or other status," and that "everyone has the right to life, liberty, and security of person." And although the Declaration never asserts a divine origin for these human rights, it does maintain that they are "equal and inalienable rights of all members of the human family," which means that all individuals of the human species are in possession of them and that they cannot be given or taken away.

Many assert that this respect for human life is due only to persons, but if we examine the literature, we discover that there are many different and conflicting notions as to what constitutes "personhood." This is not surprising in an age that eschews metaphysics and asserts that much of our understanding of reality is invented or created or is the product of interpretation. I suggest that a more robust respect for human life would emerge if we were to acknowledge that all individuals of the human species are in possession of these "equal and inalienable rights" and that organisms on their way towards human individuation should be accorded some measure of respect and dignity in light of what they are destined to become—human individuals, some of whom will develop into full personhood, whatever "personhood" means.

Since *Roe v. Wade* (1973)[15] legalized abortion in the United States, many philosophical and theological battles have been fought over the concept of personhood. Germain Grisez[16] equates personhood with a human genotype, and personhood is sometimes considered the equivalent of an entity that is animated with a spiritual soul (Aquinas).[17] Conditions for personhood have been suggested, and H. Tristram Engelhardt has introduced the concept of "social" personhood for situations where the conditions for personhood don't seem to be met but where individuals or society wish to endow the "person" in question with human rights.[18]

All of these approaches seem to consider "personhood" as a metaphysical problem to be resolved. The fact that after more than 25 years we still have not developed a consensus on the attributes of personhood should indicate that we are dealing with an issue analogous to the Derridean quest for justice, a "passion for the impossible," as described by Caputo,[19] which must nonetheless be sought with full realization that our conceptual frameworks are inadequate to the task and will always provide a "remainder," a "something more" that needs to be said. My reflections here are therefore not intended to be "metaphysical" as I am not about to attempt to develop a philosophical theory of personhood. The development of such a theory is exactly what has proven difficult, if not impossible, to do, as has been amply illustrated by a recent issue of the *Kennedy Institute of Ethics Journal*.[20] Rather, I am going to begin with the assertion, held by many as intuitive, that a person (in a nonmetaphysical sense), whatever and whoever it is, is an individual of a living species, some of whose members demonstrate the capacities of thinking, choosing, and reasoning, and that this individual (person) is the bearer of inalienable rights. Thus, individuation within the human species is both a necessary and sufficient condition for personhood (loosely construed), and the fact of human individuation is enough to claim fundamental human rights. Therefore, the question of human rights, including the right to life, should focus on human individuation; that is, when does human biological life become a human individual. My reflections on personhood below will assume this "rights-bearing" approach and not some more metaphysically developed concept of personhood.

Boethius describes personhood as "an individual substance of rational nature."[21] Several comments are in order: first, even disregarding the metaphysics of substance, Boethius holds that to be a person one must be an individual. That is, until an entity can be this and only this human individual it is not a person; for Boethius, human individuation is a necessary condition of personhood. The fact that he describes it as "an individual substance" might be confusing unless we understand that Boethius is using the word "substance" here not to designate an essence (metaphysical personhood) but rather, much as Aristotle did when he spoke of "primary" or "first" substance, that which is always singular and concrete. For Boethius a person is an existential "who" not an essential "what." The person is always a substantial individual with a rational nature. The "who" for Boethius comes from the fact that this individual entity has a rational nature. I take Boethius to mean here that this individual belongs to a species, some of whose members experience freedom, thought, and the capacity to reason. The word nature here means that the thrust, goal, or *telos* of members of this species is toward rationality but is does not mean that humans (or other rational species) as individuals must possess rationality to be persons (that is, bearers of rights). Thus, for Boethius not only is human individuation a necessary condition for personhood and its rights, but it is a sufficient condition as well. What are the implications of this description? First, I think that Boethius would wish to count as bearers of rights those humans who are demented, either as a result of genetic or developmental anomalies, or as a result of injury or disease. The fact that they are individuals of a species, some of whose members are rational, would be enough for them to count as persons. The same holds true of fetuses, anencephalics, children, and sleeping people. However, I would argue on the basis of reflections provided in this article that an individual living organism (with a human genotype), developing into a member of the human species, is not a human individual until such time as this embryo cannot potentially become more or less than one individual of the species.

Thomas Aquinas uses Boethius' description as a basis for his own analysis and reflection. Aquinas says that persons are "subsisting beings, distinctive because of their intellectual nature."[17] By "subsisting beings" Aquinas means those who exist not in another but in themselves."[22] In addition to emphasizing Boethius' notion of individual existence, Aquinas wants to see persons as complete, not in the sense of finished products, but in the sense of being able to stand on their own. Therefore, a person is not able to be an accident or a semi-substance, but is a complete singular substance, a member of a species some individuals of which can think and choose.

What does this analysis suggest about when human individuation begins? Aquinas argues that personal being is incommunicable, and this in three ways: first, persons are incommunicable with regard to being parts of something because persons are complete in themselves,[23] that is, they are individuals with a rational nature. This aspect should raise significant questions with regard to the status of the fertilized ovum up until the point of the formation of the primitive streak. The developing cellular organism can be, and is, one unique being of the human species, no more or less, from this point on. As mentioned before, this happens about 14 days into the developmental process. Before this time, the totipotentiality of the cells of the inner mass of the very early embryo is such that they can split spontaneously, forming identical twins, or can be chemically cleaved (at an earlier stage) to form what are referred to as blastomeres. In addition, totipotential cells that have been cleaved can be fused again to form a single developing cell mass that will eventually become one human being, sometimes referred to as a chimera. Finally, two developing cell masses of totipotent cells of different genotypes can fuse and form one developing cell mass, also called a chimera.[11] At the blasto-cyst stage (a few days after fertilization), the embryonic cells in the inner cell mass can be changed by researchers from totipotent (able to form new individuals) to pluripotent (able to form many different tissue types) and can be grown in tissue culture as stem cells, which can now be stimulated to grow into many types of tissues and will eventually be modified to grow whole organs. As I understand Thomas Aquinas on this issue, if a being x can potentially be part of being y, then being x cannot possibly be a person (because not yet a unique individual of the human spe-cies) until such time as the possibility of its being part of some-thing else is closed off. Thus, I would argue that from fertilization until the formation of the primitive streak, a developing individual life with a new human genotype is present, but not a life with a rational nature, a human individual, a person. To hold otherwise would be to reject Thomas' notion of *incommunicabilitas partis*. For further thoughts on this issue and an excellent discussion of some of these phenomena, see Shannon and Wolter.[24]

Second, persons are incommunicable with regard to com-munality of being (*incommunicabilitas universalis*). That is, per-sons are unique subjects who cannot be objectified by discussing them in universal terms. If human persons cannot be captured by a concept, then each remains a mystery that we can reflect on and marvel at but cannot really understand. This pertains to the mys-tery of every individual human being, even those who seem to have minimal relational potential, and even to the mystery I am to myself. This dimension of personhood also points to the impor-tance of seeing the individual as an integrated whole, a quality that totipotent cells (on their way to becoming human individuals to be sure) do not possess completely because they can form iden-tical twins and chimeras. We cannot call identical twins by name before they are formed, and we would be foolhardy to call totipo-tent cells by name because it might be possible for them to divide and form twins and even possible for these divided cells to eventually fuse back together to form one human individual. Humans are unique and irreplaceable individuals who are incommunicable with regard to the communality of being.

Finally, Aquinas argues that human persons cannot be assumed into some higher being (*incommunicabilitas assumptibilis*) because that which is assumed crosses into the personality of the other and does not have its own personality. Although Thomas Aquinas was considering the divine personhood of Jesus Christ in this context, it seems to me that Aquinas would reject out of hand the personhood of totipotent human cells (just as he rejects the human personhood of Jesus Christ) precisely because they have the potential of being assumed into some higher being (a chimera), even though this rarely happens in fact.

But does reverence for human life demand human individuation? By no means. If we believe in the interconnectedness and interrelatedness of all creation, as process philosophy and environmental ethics would suggest, organized cells with a human genotype (pre-human individuals not yet members of the human species) that are "on their way" towards personhood should be accorded the highest reverence, dignity, and respect even though they are not in possession of the "equal and inalienable" rights of an individual human just yet. Thus, I would agree with Shannon and Wolter that "all life is a many-splendored creation," and "this is especially true of human life in any stage of its development."[24] I do hold that respect for life when human individuation has not yet occurred provides new insight for allowing some stem cell research, a topic I will discuss in detail below.

Sources of Stem Cells and the Ethical Implications of Their Derivation

Having analyzed the relationship between human rights, human dignity, and individuation, I would like to analyze the law in relationship to stem cell research. However, I would first like to conisder, in more detail, the sources of stem cells and the ethical implications of their derivation.

Stem Cells From Living Persons (or From Umbilical Cord Blood)

First of all, we need to question whether there are any ethical issues involved when deriving stem cells from living persons (or from umbilical cord blood), and if so, what are they? It seems to me that two types of ethical issues are present in this situation: issues of informed consent and issues of ownership. It is true that almost all religious bodies, and people generally, accept the derivation of stem cells from living humans (or from umbilical cord blood preserved at birth) as ethical; yet there are still safeguards that must be considered if this derivation is to be truly in accord with the principles of biomedical ethics as understood today.

The first consideration must be that of informed consent. If cells are being taken from the body to be cultured as stem cells, the person from whose body the cells will come has the right to be informed of the benefits, and possible risks, of this procedure, must understand these benefits and risks, and must freely and without coercion give permission for these cells to be removed. If the person in question is unable to understand the information or to give consent, proxy permission must be sought and granted. If this removal of cells is part of a research protocol, all the requirements relating to the protection of human subjects in research must be followed.[25]

Informed consent is generally analyzed into five components, the first being the threshold component of medical competency. The ability to make medical decisions is described as a threshold principle because any process of informed consent must begin with a determination of this ability. Often this is described as competency, but it should be emphasized that this does not refer to legal competency (usually determined by a judge), but the ability to make medical decisions (usually determined by the attending physician).[26] If the person is unable to make reasoned decisions regarding his or her own health care, then a proxy (either appointed by the patient, or an appropriately chosen family member or friend) able to make medical decisions for the patient, who acts in the name of the patient, or knows generally what the patient would want, or is able to make a decision in the best interest of the patient, can give permission for medical treatment or therapeutic research to continue. This is not, strictly speaking, informed consent, but rather "informed permission," if the other four elements described below are met.

The second and third elements refer to the "informed" part of informed consent. The second is disclosure of information. The physician must disclose to the patient any aspects of the treatment at hand (here the removal of stem cells) that a reasonable person would want to know in order to make a reasoned decision. This would include benefits to the patient and others: costs, risks, possible harms, and any other information relevant to this specific case. Ideally, the physician should tell the patient everything the patient wishes to know regarding the treatment, even if it is more than a reasonable person generally would want to know. The third element is understanding. Not only must the physician present information relevant to the situation at hand, the physician has an obligation to do everything possible to assure that the patient understands the information the physician is providing. This should be undertaken over a period of time, with follow-ups by nurses, administration of questionnaires, and so on. All that is demanded in this regard is an effort of good faith on the part of the physician, since there can be no guarantee that patients really understand all the intricacies of their medical situation.

The fourth and fifth elements deal with consent. The fourth states that the patient must not be coerced by force or fear into making a decision by the physician, family, or well-meaning friends. That is, medical decision making, even about the removal of stem cells, must be free and unforced. This freedom should also include freedom from fear, an ideal that can rarely be perfectly met in the clinical setting. Finally, the fifth element of informed consent is the actual acceptance or rejection of the proposed medical intervention by the patient or proxy. This is usually communicated by a written document, "the informed consent," but it can be communicated verbally as well.

The other ethical issues that must be considered, when physicians request removal of stem cells from adults or permission to freeze umbilical cord blood for possible use in developing stem cells later on, are those of ownership. Ownership issues can generally be settled contractually before any procedure takes place, but what must be dealt with is the possible patenting of the stem cells after their removal. (Current law is fairly clear on both these issues but what I am raising here are ethical concerns that are less clear.) To be eligible to receive a patent, an invention must be of "patentable subject matter." The patent system is not designed to reward an inventor for the act of discovering something that was present in the natural world yet undiscovered before the inventor came along. The system is designed to reward an inventor for developing a new and useful product as a result of his time and energy expended. In *Diamond v. Chakrabarty* (1980), the Supreme Court of the United States held that living organisms were not automatically discounted as unpatentable subject matter.[27] This decision is credited with enabling the rapid development of the biotechnology industry in the United States.[28] In 1987, the Commissioner of Patents of the Patent and Trademark Office presented an official policy announcement stating that, "the Patent and Trademark Office now considers non-naturally occurring nonhuman multicellular living organisms, including animals, to be patentable subject matter.[29] It remains to be seen whether the development of unique stem cell lines derived from adults or umbilical cord blood will be granted patents, but if that is to be the case, potential donors should at least be informed of the fact before any contracts outlining ownership are drawn up.

In fact, as was well stated by a California court, "If science has become science for profit, then we fail to see any justification for excluding the patient from participation in these profits" (*Moore vs Regents of Univ. of Cal.*, 1988).[30] However, it should also be pointed out that the Supreme Court of California, reversing this judgment in part, held that once cells are excised from the body, they are not the property of the person from whom they have been excised.[31] I believe this decision will be rightly challenged the more bodily parts, including stem cells, are used for potential commercial benefit. Even in the face of current law, physicians should deal with this issue and discuss it with stem cell donors before removing stem cells, especially when they remove them for research purposes.

Except for the issue of informed consent and the possible issues of research involving human subjects and patenting and ownership decisions, there do not seem to be other ethical issues affecting the decision to remove adult stem cells or to save for possible future stem cell derivation the blood cells of the umbilical cord at birth. Adult stem cell research is generally considered a beneficent activity; many religious bodies accept it; even the pro-Life Committee of the United States Conference of Catholic Bishops supports it.[32] Because this is so, and because research involving adult stem cells has shown such promising results, it should be high on the list of priorities for federal funding for biomedical research, if such research is to be funded at all (*see* the section on *Issues of Justice* above).

Stem Cells From Frozen Embryos

Support for stem cell research is greatly weakened when the stem cells are derived from frozen embryos. These embryos are the product of in vitro fertilization and are produced as "spare embryos" to be used if implantation of one or more embryos do not attach during the first or subsequent rounds of embryo transfer to the uterus of the woman. When these spare embryos are not used, they are generally thawed (they die during this process) and are discarded. What, then, are the ethical issues involved in using stem cells derived from frozen embryos for research, and what should the position of the federal government be in terms of supporting this research?

The first set of ethical issues would be the same as those discussed previously with regard to stem cells derived from living persons or from umbilical cord blood. Informed consent would have to be obtained from the genetic parents of the embryo, and a discussion and agreement about ownership and patenting of any future stem cell line would have to be made in accordance with the law.

But the major and most divisive ethical issue in using frozen embryos as sources for stem cells turns on the moral status of the developing embryo. There are many who would argue that even if the genetic parents give permission, spare embryos should not be used as sources for stem cells because these frozen embryos are individuals of the human species and the bearers of rights, specifically the right to life as an individual. People from many religious traditions (and others who are not religious) hold this view, but since the Catholic tradition explains this view very clearly, I will rely on Catholic teaching to explicate this stance. As Cataldo points out, the Catholic tradition acknowledges that its teaching "on the beginning of human life has always deferred to the established scientific knowledge on the question."[33] This probably will come as a surprise to many people who think that the Church's position on this issue is dogmatically based on faith. Nonetheless, Cataldo continues, "on the basis of the latest scientific data, the Church concludes that from the moment the human zygote is formed a new human individual exists."[33] This point has been explicitly made by the Church many times since 1974. Cataldo deals with the issue of twinning (but not chimera formation—a more difficult philosophical problem it seems to me) thus: "what makes a composite being an actual individual is that its parts function primarily for the good of the whole. The embryo's potential to become another embryo is certainly there, but it does not affect its actual state as an individual."[33] It is at this point that I would like to bring in my own analysis, discussed more thoroughly above. Although I agree that all human life, especially developing human life, deserves respect; nevertheless, I think that science can point out only that an individual organism is in the process of development and cannot say (because it would be beyond its competence) that this is a human organism. Science can tell us that the zygote has a unique human genome (but creating stem cells retains this genome, so the genome itself cannot be a sufficient condition for human individuation) but the zygote and the developing embryo also have the potential for several days to be more than one human individual (twins) or less than one human individual (a chimera). This is the point at which the Catholic Church should reflect on its own rich tradition dealing with personhood and individuation, which I have presented above. Doing so might help with the insight that this life, although it should be reverenced and respected, is not the bearer of rights because it is not yet an individual of the human species because it can be potentially more than or fewer than one individual human.

This is extremely important, because, if it can be agreed that this individual organism is not the bearer of inalienable rights, then it would seem that the lesser of two evils (if evils they are), is the one that accords more respect for human life understood generically, that is, the option that at least allows the genome to live on (as a tissue culture of stem cells), not the one that allows death by thawing. Of course, there are those who argue that IVF clinics should not be producing spare embryos in the first place. I believe there is merit to these arguments if we are considering seriously the respect due to all developing human life; however, the fact is that these embryos do exist and, rather than allowing them to die by thawing, we should be able to use them for developing new stem cell lines which may have significant therapeutic value. Ironically, because the genome of these cells is maintained, it is theoretically possible that one of these cells might be cloned to produce actual individuals with the genome, something suggested as theoretically possible by McGee and Caplan a few years ago.[34] It seems to me, then, that the derivation of embryonic stem cells from frozen embryos (which have been frozen in the context of in vitro fertilization) can, in some instances, be seen as the lesser of two evils, and also as the approach that respects life more genuinely than does thawing, and on that account be considered ethical. Whether or not the federal government should fund this type of activity is another question.

Creating Embryos for Stem Cells

The creation of embryos with the specific intention of deriving stem cells from them is ethically problematic whether the embryo is produced by in vitro fertilization, cloning, or parthenogenesis. There is general consensus among ethicists that humans should never be used as a means only, but always as ends in themselves. Though I have argued that the developing embryo is not, strictly speaking, a human until the primitive streak forms and cell differentiation takes place, it nevertheless seems appropriate that embryos (which are at least on their way towards becoming human individuals) should be accorded the same respect and not be created with the intention of destroying them (as embryos) in a few days. Reverence for human life must entail respect for the developmental process by which humans are formed. Even deriving stem cells from the embryonic cell mass of embryos frozen for in vitro fertilization, which are not now going to be used, shows them some respect, because, as I have argued above, it is a better alternative than death. This argument based on the reverence and respect due to developing human life, although not as powerful as the one which asserts that these entities possess inalienable rights—including the right to life, should be enough to ground ethical opposition to the creation of embryos solely for the purpose of stem cell research. There is widespread ethical and political consensus for this point of view.[35] However, there may be another approach to deriving stem cells from egg cells that would be less controversial. Experiments are now being conducted which show "that monkey eggs can be chemically treated and modified to the point where they begin behaving enough like embryos to generate stem cells—all without the addition of sperm normally required for embryogenesis, and without any capacity to grow into a baby monkey."[36] If research using parthenogenesis in humans could produce pluripotential stem cells directly and not totipotential embryonic stem cells first, there would be little ethical debate because at no time would a developing (or for many, an actual) human individual be formed. This could be a source of stem cells that have all the potentially good qualities of embryonic stem cells and have the ethical advantage of not having been derived from an embryonic cell mass.

Ethics and the Law

For many years the federal government has refused to fund research on the fetus unless there is the likelihood that the fetus itself will be directly benefited by this research (therapeutic research). On October 21, 1998, Congress passed the Omnibus Consolidated and Emergency Supplemental Appropriations Act of 1998 (OCESAA).[37] This law extends the reach of 45 CFR 46 and prohibits the use of federal funds for research involving human embryos as well as fetuses. Human embryo is defined in section 511 (b) of the act as: "any organism, not protected as a human subject under 45 CFR 46 as of the date of enactment of this Act, that is derived by fertilization, parthenogenesis, cloning, or any other means from one or more human gametes or human diploid cells." In effect, this means that the federal government will not fund research designed to create embryos for the purpose of deriving stem cells from them.

After pluripotent stem cell lines had been derived from human embryos through private funding sources, Dr. Harold Varmus, at that time Director of the NIH, asked the Department of Health and Human Services (DHHS) general counsel, Harriet Raab,[38] to prepare a legal opinion. Specifically, Varmus asked Raab if federal funds could be legally used for pluripotent stem cell research that uses a cell line derived from aborted fetuses or embryos from IVF clinics. Raab responded that federal law permits the NIH to support such research, basing her opinion on an attenuated interpretation of OCESAA. Raab stated that "[p]luripotent stem cells are not a human 'organism' as that term is used in the definition of human embryos provided by the statute" and that human pluripotent cells are not "even precursors to human organisms."[38]

In response to the DHHS opinion, the NIH issued a moratorium on federal funding of pluripotent stem cell research until it could develop guidelines. In April 1999, the NIH formed an Advisory Committee to create these guidelines and to develop oversight rules for this research. This group met in public session and included scientists, clinicians, ethicists, lawyers, patients, and patient advocates. The group also received input from the National Bioethics Advisory Committee.[39] The final version of the guidelines was published on November 21, 2000.[40] The guidelines state that the embryos must have been created completely without federal funding and, to ensure that the donation of embryos is voluntary, state that no inducements may be offered for the donation of the embryos. Strict rules for informed consent of potential donors of embryonic stem cells are mandated by the guidelines, but they also state that "unlike pluripotent stem cells derived from human embryos, DHHS funds may be used to support research to derive pluripotent stem cells from fetal tissue, as well as for research utilizing such cells," as long as the abortion occurs independently of the research and the fetus is donated without coercion.[40]

During 2001 there was much discussion in Congress about stem cell research, both embryonic and adult, as well as discussion of "therapeutic cloning," that is, somatic cell nuclear transfer with the intention of deriving stem cells from the embryonic clone. To date, however, there has been no congressional legislation on any of these issues. Most of the discussion in Congress focuses on issues I have already presented, and, since no legislation has yet been passed, I will provide just a few examples of positions taken by members of the House and Senate.

On June 19, 2001, Senator Sam Brownback (R-KS) discussed federal funding of both embryonic stem cell research and cloning and said that the two were "inexplicably tied together."[41] His position is that "federally funded human embryonic stem cell research is illegal, immoral, and unnecessary for where we are and what we know today."[41] And because he opposes embryonic stem cell research in general, it is clear that he also opposes cloning to derive stem cells. He applauds President Bush, who at that time was on record as opposing "federally funded research for experimentation on embryonic stem cells that require live human embryos to be discarded or destroyed."[41]

About a month later, on July 18, 2001, the only physician in the Senate (also a transplant surgeon and now Senate Majority Leader), William Frist (R-TN),[42] presented a very thorough overview of the whole stem cell debate, emphasizing that it is not only a scientific or medical issue, but an ethical and political one as well. Because these points presented by Senator Frist are so important (and so sensible), I will present them verbatim:

> There are basically 10 points I think we must consider, and I have proposed an answer. Again, I don't know the answer, and I struggle, like every person, on this particular issue to make sure we have the appropriate moral considerations. But I will outline what my 10 points are.

> No. 1, we should ban embryo creation for research. The creation of human embryos solely for research purposes should be strictly prohibited.

No. 2, we should continue the funding ban on the derivation of embryonic stem cells. We need to accomplish this by strengthening and codifying the current ban on federal funding for the derivation of embryonic stem cells.

No. 3, we should ban human cloning. We need to prohibit all human cloning to prevent the creation and the exploitation of life for research purposes.

No. 4, we should increase adult stem cell research funding. These adult stem cells, stem cells that are removed from an adult, that you can back out in such a way that you can capture the potential for using them for treatments for various diseases—we should increase this funding for research on adult stem cells to ensure the pursuit of all promising areas of stem cell research, on both adult stem cells which occur much later in life and the embryonic stem cells which are derived at the 5- or 6-day-old blastocyst stage.

No. 5, provide funding for embryonic stem cell research only from blastocysts that would otherwise be discarded. We need to allow federal funding for research using only those embryonic stem cells derived from blastocysts that are left over after in vitro fertilization and would otherwise be discarded.

No. 6, require a rigorous informed consent process to ensure that the blastocysts used for stem cell research are only those that would otherwise be discarded. We must require a comprehensive informed consent process establishing a clear separation between a potential donor's primary decision to donate blastocysts for adoption or to discard blastocysts and their subsequent option to donate blastocysts for research purposes. Such a process is modeled on the well established and broadly accepted organ and tissue donation process in which I have been so intimately involved over the last 20 years.

No. 7, limit the number of stem cell lines. I believe we should restrict federally funded research using embryonic stem cells derived from blastocysts to a limited number of cell lines. This does not mean limiting it to research using stem cells that have already been derived to date, most of which would reportedly not be eligible even under the current NIH guidelines that need much strengthening. In transplantation, when I remove a heart from an individual and I give it to another individual, that one individual benefits. With stem cells, it is very different. From a stem cell line, you derive the cells, and that stem cell line can be used for multiple experiments, thousands of investigations, as we go forward.

No. 8, establish a strong public research oversight system. I believe we should establish an appropriate public oversight mechanism, including a national research registry, to ensure the transparent, in-depth monitoring of federally funded and federally regulated stem cell research and to promote high ethical, moral, and quality research standards.

No. 9, require ongoing, independent scientific and ethical review. We need to establish an ongoing scientific review of stem cell research by the Institute of Medicine and create an independent Presidential advisory panel to monitor evolving bioethical issues in the area of stem cell research. In addition, we need to require the Secretary of Health and Human Services to report to Congress annually on the status of federal grants for stem cell research, the number of stem cell lines created, the results of stem cell research, the number of grant applications received and awarded, and the amount of federal funding provided.

Lastly, No. 10, strengthen and harmonize fetal tissue research restrictions. Because stem cell research would be subject to new, stringent federal requirements, I believe we must ensure that informed consent and oversight regulations applicable to federally funded fetal tissue research be made consistent with these new rules.

Senator Frist in this statement describes himself as pro-life; it seems to me that this description is accurate. However, he himself comes down on this issue almost exactly where I do for some of the same reasons I have discussed above.

The "Human Cloning Prohibition Act of 2001" was passed by the House of Representatives on July 31, 2001.[43] This bill "does not in any way impede or prohibit stem cell research that does not require cloned human embryos." "It does prohibit the creation of cloned embryos." In the debate leading up to the approval of this bill, it was pointed out that the bill would not only prohibit human cloning, but also any attempt at human somatic cell transfer, even for the purpose of producing stem cells. Thus the Greenwood/Deutch amendment was proposed that would ban the production of human clones, but would allow human somatic cell transfer for the purpose of producing embryonic stem cells. But the counterargument was that to get the stem cells, an embryo must be produced, an embryo that could possibly not be used for stem cells but rather be allowed to grow as a cloned human being. The Greenwood/Deutch amendment was defeated and the bill as written passed the House by a 265-162-6 majority. The previous Senate did not act on this bill.

On November 2, 2001, Ruth L. Kirschstein, Acting Director, National Institutes of Health, revoked the existing NIH (2000) guidelines insofar as they deal with stem cells derived from human embryos.[44] The reason for this revocation was the decision made by President Bush on August 9, 2001, despite the opinion of General Counsel Raab, to limit funding of embryonic stem cells to those cell lines already in existence before that date.[45] This decision effectively rendered moot many of the provisions of the NIH guidelines discussed earlier because the federal government will no longer fund research on embryonic stem cell lines derived from frozen embryos produced by in vitro fertilization or by cloning after August 9, 2001. This is the legal situation in relation to embryonic stem cells as of February 2003.

However, on September 5, 2001, a little less than a month after President Bush's decision, several prominent Senators from his own party wrote a letter to the president asking him to rethink his prohibition.[46] Their argument was that if embryos were going to be destroyed anyway, it would be better if embryonic cell lines were produced from them, and if government funding were provided both to derive these embryonic stem cells and to do research on these new cell lines once they are established. Shortly after this, the United States endured the horror of the terrorist attacks of September 11, 2001 and many of these issues were put on the back burner. They resurfaced again with the advent of the President's Advisory Commission established by President Bush when he made his announcement about federal funding for research using already established stem cell lines. In July 2002, the Council concluded unanimously that reproductive cloning should not be permitted, and in a 10-to-7 decision endorsed a moratorium on research cloning for 4 years.[47,48]

Final Observations

The aim of this chapter has been to review the literature dealing with stem cell research and its relation to distributive justice of health care resources, to the moral status of the embryo, to how stem cells are derived and the ethical implications of these procedures, and to the ethical debate underlying the issues of public funding and the development of the law in this area. But I would like to close by considering some caveats in relation to stem cell research and some recommendations regarding federal funding of stem cell research that flow from my previous analyses. My first observation is that in dealing with stem cell research, we must beware of "hype." Ten years ago scientists were telling us that if adequate funding were provided to decode the human genome, many medical benefits would follow in its wake. The fact is we have now decoded the genome and relatively few medical applications have been developed that have proven successful. The same kinds of promises (and requests for funding) are being made for embryonic stem cell research,[49] and, although it is hoped that some of them will prove true, there has yet to be a clinical application of embryonic stem cell research that has proved beneficial to humans. Stem cell research, generally, and embryonic stem cell research specifically, are in a very early stage of development, and it is unlikely, despite claims to the contrary, that stem cells will prove to be "magic bullets" that will quickly remedy humanity's ills. It is likely that effective treatments for some conditions will be developed, but that this will happen very much in the future. To his credit, Senator Frist pointed this out to his colleagues thus:

> However, it is important that advocates not over-sell the potential of either embryonic or adult stem cell research for medical treatments. This evolving science is relatively new, and much basic research remains before we can reasonably expect to see clinical trials and possible treatments. In fact, to date, with the exception of hematopoietic stem cells that have been used in bone marrow transplantation for many years, none of these sources has yet demonstrated proven therapeutic applications.[42]

Second, we must not let the controversy surrounding embryonic stem cell research deter us from funding research using adult stem cells and stored blood cells from the umbilical cord. Some research using these stem cells has already proven successful and, since there is little ethical debate regarding their use, both the federal government (if the norms for distributive justice that I have discussed can be met) and other entities who oppose embryonic stem cell research (e.g., the Catholic Church) should be in the forefront of supporting this area of proven clinical success.

Finally, I would support President Bush's approval of funding existing stem cell lines for the reasons he cites in the press release of August 9, 2001, but I would go further than the president and approve funding of research using newly created cell lines from frozen embryos intended originally for implantation but now destined to be thawed and destroyed. I have explained my reasoning for this by arguing that it seems to me the lesser of two evils to allow stem cell derivation from those frozen embryos that would otherwise be destroyed. However, I would minimize government funding of the creation of these stem cell lines on prudential more than on principled grounds, because there is a great deal of political and religious opposition to the destruction of embryos whatever their source. It seems to me that, at the least, the original NIH safeguards that were developed in this regard should be reinstated and implemented.

I would be strongly opposed to the funding of the creation (by whatever technology) of embryos with a human genome with the intention of using them to derive embryonic stem cells. This process devalues us as humans even as it devalues the embryonic processes that would eventually lead to human individuation. The sacredness and sanctity of human life should extend beyond the rights of individuals and encompass all those processes that lead to individuated human beings. And only if the interruption of this process is the lesser of two evils (in the case of frozen embryos created for implantation but now destined for destruction) can I find it ethically justifiable, even though I would minimize government funding for it as I have explained above.

One thing is very clear at this point: we have only begun to delve into the ethical, political, and legal conundrums that contemporary biological science and technology have thrust upon us. It is my hope that this chapter will help to stimulate that debate.

Notes and References

[1]Fins, J. J. and Schachter, M. (2002) Patently controversial: markets, morals and the president's proposals for embryonic stem cell research, *Kennedy Institute of Ethics Journal*, **12(3)**, 265–278.

[2]Hansen, J-E. S. (2002) Embryonic stem cell production through therapeutic cloning has fewer ethical problems than stem cell harvest from surplus IVF embryos, *Journal of Medical Ethics*, **(28)2**, 86–88.

[3]Lagnado, L. (2002, November 12). Uninsured and ill, a woman is forced to ration her care, *The Wall Street Journal*. pp. A1, A14.

[4]Center for Clinical Bioethics, Georgetown University Medical School (2002). Conference: Just health care: moral critique, outrage and response. (April 10–12, 2002).

[5]Hackler, C. (2001) Justice and human nature, *American Journal of Bioethics* **(1)2**, 38–39.

[6]Holland, S., Lebacqz, K., and Zoloth, L. (eds.). (2001) *The Human Embryonic Stem Cell Debate: Science, Ethics and Public Policy*. MIT Press, Cambridge, MA.

[7]The Stem Cell Debate. *The New York Times*, December 18, 2001, Section F.

[8]Hall, S. S. (2001) Adult stem cells. *Technology Review*, November. Accessed from the Internet on December 12, 2001 at http:// www.techreview.com/magazine/nov01/hall.asp

[9]Kaiser Daily Reproductive Health Report (2001). Private cord blood bank opens to criticism in Australia, (December 19, 2001). This article reports the following: "A private umbilical cord blood bank owned by biotechnology company Cryosite opened Monday in Sydney, Australia, to criticism from supporters of the publicly funded National Cord Blood Bank in Melbourne, the Age reports. Umbilical cord blood is a significant source of blood stem cells, which are capable of replenishing bone marrow, regenerating blood-forming cells and restoring immune function in people with blood disorders and diseases like leukemia. Parents may choose to store their infant's cord blood at birth, either publicly or privately, in case their child or another family member later becomes ill and needs stem cells that are a perfect genetic match. Cord blood stored in a private bank can be accessed only by the

donating family, whereas public banks allow the frozen cells to be used by others in need of a close match. Dr. Simon Bol, scientific director of the public bank at Melbourne's Royal Children's Hospital, said that 'anybody in Australia can search for a cord blood match' at his institution. The new private bank charges parents an initial collection fee of about $1,000 and a yearly storage fee of about $80. The public bank, which is backed by a $4.6 million government grant, seeks to collect 22,000 donated cord blood units for public use by 2004." Accessed from the Internet on December 21, 2001: http://www.kaisernetwork.org/daily_reports/rep_index.cfm?DR_ID=8617

[10]Milde, A., Kuhl-Burmeister, R., Ritz-Timme, S., and Kaatsch, H. J. (1999) DNA typing in cases of blood chimerism, *International Journal of Legal Medicine* **112(5)**, 333–335.

[11]Strain, L., Dean, J.C.S., Hamilton, M. P. R., and Bonthron, D. P. (1998) A true hermaphrodite chimera resulting from embryo amalgamation after in vitro fertilization, *The New England Journal of Medicine* **338(3)**, 166–169.

[12]Ford, N. (1988) When did I begin: conception of the human individual in history, in *Philosophy And Science*. Cambridge University Press, New York.

[13]Kavanaugh, J. F. (2001) *Who Count As Persons?: Human Identity And The Ethics Of Killing*. Georgetown University Press, Washington, DC.

[14]General Assembly Of The United Nations (1948) *Universal Declaration Of Human Rights*. Accessed from the Internet on December 5, 2002, http://www.un.org/Overview/rights.html

[15]*Roe v. Wade*. (1973). **410** U.S. 113. Science and morality: no conflict. *Life Issues Forum*, (August 18). Accessed from the Internet at http://www.nccbuscc.org/prolife/publicat/lifeissues/08182000.htm

[16]Grisez, G. (1970) *Abortion: The Myths, The Realities, and The Arguments*. Corpus Books, New York.

[17]Aquinas, T. (13th Century CE, a) *De Potentia Dei*.

[18]Engelhardt, H. T. (1973) The beginnings of personhood: philosophical considerations, *Perkins Journal*, **27**, 24.

[19]Caputo, J. (1997) *The Prayers and Tears of Jacques Derrida: Religion Without Religion*. Indiana University Press, Bloomington.

[20]*Kennedy Institute of Ethics Journal* (December 1999).

[21]Boethius, A. M. S. (6th Century CE). *Liber de Persona et Duabus Naturis in Boetii Opera Omnia, Vol II.* Patrologiae Latinae Tomus 64—Migne.

[22]Aquinas, T. *Summa Theologica I.*

[23]Aquinas, T. *Scriptum Super Sentientiis, Tomus III.*

[24]Shannon, T. A. and Wolter, A. B. (1990) Reflections on the moral status of the pre-embryo, *Theological Studies*, **51**, 603–626.

[25]45 Code of Federal Regulations (CFR) 46 (2002) Federal guidelines relating to human subjects' research.

[26]Beauchamp, T. and Childress, J. (2001) *Principles of Biomedical Ethics, Fifth Edition.* Oxford University Press, New York.

[27]*Diamond v. Chakrabarty*, (1980). 447 U.S. 303 at 311.

[28]Newman, P. (1997) Intellectual property law and the new biology, *Judges' Journal*, **36(3)**, 46.

[29]Office of Technology Assessment (OTA), U.S. Congress (1989) *New Developments in Biotechnology: Patenting Life—Special Report.* Pub. No. OTA-BA-370, at 93.

[30]*Moore v. Regents of Univ. of Cal.*, (1998). **249** Cal. Rptr. 508 (Cal. Dist. Ct. App.).

[31]*Moore v. Regents of University of California* (1999) **51** Cal. 3d 120 at 141.

[32]Doerflinger, R. M. (2000) for the United States Conference of Catholic Bishops.

[33]Cataldo, P. J. (2001) Human rights and the human embryo, *Ethics and Medics* **26(12)**, 1–2.

[34]McGee, G. and Caplan, A. (1999) The ethics and politics of small sacrifices in stem cell research, *Kennedy Institute Journal of Ethics* **9(2)**, 151–165.

[35]Stolberg, S. G. (2001) Controversy reignites over stem cells and clones, *The New York Times*(December 18), F1. See F8 in the same paper for some of the groups being gathered to oppose creation of embryos by cloning.

[36]Wolf, D. P. (2001) Parthenogenetic activation of rhesus monkey oocytes and reconstructed embryos, *Biology and Reproduction* **65**, 1253–1259.

[37]US Congress. (1998) Omnibus consolidated and emergency supplemental appropriations act of 1998. *P.L.* **105–277, 112 Stat.** 2681.

[38]Raab, H. S. (1999) Federal funding for research involving human

pluripotent cells, *Biolaw* **2S,** S:60–61.

[39]NBAC (1999) *Ethical Issues In Human Stem Cell Research, Volume I: Report And Recommendations Of The National Bioethics Advisory Commission.*

[40]NIH (2000) *National Institutes of Health Guidelines for Research Using Human Pluripotent Stem Cells. Federal Register* **65,** 51976, (August 25, 2000), corrected Federal Register **65,** 69951, (November 21, 2000).

[41]Brownback, S. (2001) *Congressional Record,* S6393.

[42]Frist, W. (2001) *Congressional Record,* S7846–S7851.

[43]House of Representatives (2001) Human cloning prohibition act of 2001. *Congressional Record,* H4916–H4945.

[44]This is the text of the document as it appeared in the FEDERAL REGISTER on November 2. 2001: DEPARTMENT OF HEALTH AND HUMAN SERVICES

National Institutes of Health, National Institutes of Health Guidelines for Research Using Human Pluripotent Stem Cells

ACTION: Notice; withdrawal of NIH Guidelines for Research Using Pluripotent Stem Cells Derived from Human Embryos (published August 25, 2000, 65 FR 51976, corrected November 21, 2000, 65 FR 69951).

SUMMARY: The National Institutes of Health (NIH) announces the withdrawal of those sections of the NIH Guidelines for Research Using Human Pluripotent Stem Cells, http://www.nih.gov/news/stemcell/stemcellguidelines.htm. (NIH Guidelines), that pertain to research involving human pluripotent stem cells derived from human embryos that are the result of in vitro fertilization, are in excess of clinical need, and have not reached the stage at which the mesoderm is formed.

The President has determined the criteria that allow federal funding for research using existing embryonic stem cell lines, http://www.whitehouse.gov/news/releases/2001/08/print/20010809-1.html. Thus, the NIH Guidelines as they relate to human pluripotent stem cells derived from human embryos are no longer needed.

FOR FURTHER INFORMATION CONTACT: NIH Office of Extramural Research, NIH, 1 Center Drive, MSC 0152, Building

1, Room 146, Bethesda, MD 20892, or e-mail DDER@nih.gov.

[45]*See* the Press Release of August 8, 2001 on this subject. Press release accessed from the Internet on December 28, 2001 at http://www.whitehouse.gov/news/releases/2001/08/print/20010809-1.html. This release states the following:
Embryonic Stem Cell Research
August 9, 2001
"As a result of private research, more than 60 genetically diverse stem cell lines already exist." I have concluded that we should allow federal funds to be used for research on these existing stem cell lines "where the life and death decision has already been made." This allows us to explore the promise and potential of stem cell research "without crossing a fundamental moral line by providing taxpayer funding that would sanction or encourage further destruction of human embryos that have at least the potential for life."
—George W. Bush

[46]Specter, A. et al. (2001) *Congressional Record*, S9118.

[47] The web site of the Council (www.bioethics.gov) provides information on all the Council's interests, including stem cells, and also has a link to the full text of its cloning report.

[48]Lyon, A. (2002) The cloning report: left of Bush but still a ban, *Hastings Center Report* **32(5)**, 7.

[49]Okarma, T. B. (2001) The technology and its medical applications, in *The Human Embryonic Stem Cell Debate: Science, Ethics, and Public Policy.*(Holland et al., eds.), MIT Press, Cambridge, MA.

Abstract

In addition to the common ethical concerns raised by any new biomedical research application, particular ethical concerns have been raised about the sources of human stem cells, especially those stems cells derived from human embryos and aborted fetuses. Catholic health care organizations share these concerns. Their affiliation with the Catholic Church prohibits them from creating or destroying human embryos or from using aborted fetal tissue for research or therapeutic purposes, and thus they cannot licitly participate in activities that involve the derivation of human embryonic or fetal stem cells. Still under consideration in Catholic health care circles, however, is the question of whether these organizations can licitly use the human embryonic and fetal stem cell lines derived by others. To date, the Catholic Church has not definitively answered this question. However, it is a question that is being vigorously debated and will soon need to be resolved. Here, I summarize the debate thus far and then try to clarify and extend it by considering the recent work of two legal scholars, M. Cathleen Kaveny and Christopher Kutz. I conclude that, should they use the embryonic and fetal stem cell lines derived by others, Catholic health care organizations cannot avoid being morally complicit in the processes by which the stem cells were derived. But I also suggest some conditions that, if addressed, might nevertheless permit Catholic agents licitly, if reluctantly, to use embryonic and fetal stem cell lines.

Complicity in Embryonic and Fetal Stem Cell Research and Applications

Exploring and Extending Catholic Responses

Jan C. Heller

Introduction

In addition to the common ethical concerns raised by any new biomedical research application (usually discussed under the headings of safety, efficacy, access, and cost effectiveness), particular ethical concerns have been raised about the sources of human stem cells, especially those stems cells derived from human embryos and aborted fetuses.[1] Catholic health care organizations in particular share these concerns. Their affiliation with the Catholic Church prohibits them from creating or destroying human embryos or from using aborted fetal tissue for research or

From: *Biomedical Ethics Reviews: Stem Cell Research*
Edited by: J. M. Humber and R. F. Almeder © Humana Press Inc., Totowa, NJ

therapeutic purposes,[2] thus they cannot licitly participate in activities that involve the derivation of human embryonic or fetal stem cells.[3] The derivation of human embryonic stem cells is regarded as the moral equivalent of abortion because the processes kill the embryo, and the derivation of stem cells from an aborted fetus is prohibited for prudential reasons having to do with the Catholic Church's public policy position on abortion.

Still under consideration in Catholic health care circles, however, is the question of whether these organizations can licitly use the human embryonic and fetal stem cell lines derived by others. This is an important question for the religious sponsors and leaders of Catholic health care, for they may face a profound moral conflict should the applications of human embryonic and fetal stem cell research be proven safe, efficacious, and cost effective. On the one hand, they will want to comply with official church teaching should the Catholic Church prohibit the use of these stem cell lines (which, as this is being written, seems likely), while, on the other, they will want to offer the therapeutic benefits of stem cell applications to their patients. Moreover, from a self-interested perspective, they will want to avoid being placed in an adverse competitive position should non-Catholic health care organizations be able to offer such applications which they could not.[4]

To date, the Catholic Church has not definitively answered the question of whether the health care organizations sponsored by its religious orders and dioceses can participate in stem cell research or offer stem cell applications when the stem cell lines in question were derived by others from prohibited sources.[5] However, it is a question that is being vigorously debated and will soon need to be resolved. Below, I summarize the debate thus far and then try to clarify and extend it by considering the recent work of two legal scholars, M. Cathleen Kaveny and Christopher Kutz.

Preliminary Catholic Appraisals

Catholic ethicists and moral theologians have produced a number of preliminary appraisals on the question under consideration here, appealing mainly to the principles governing material cooperation and to claims about complicity in wrongdoing. I explore two of these appraisals here. Both distinguish between those processes used to derive human stem cells and their use for research or therapeutic purposes, and both argue that Catholic health care organizations cannot licitly participate in the derivation of human embryonic and fetal stem cells.[6]

The Rev. Kevin D. O'Rourke employs the principles governing cooperation to ask whether it could be permissible for Catholic agents to use the embryonic and fetal stem cell lines derived by non-Catholic agents.[7] These principles may not be familiar to readers. They were developed centuries ago by Catholic moral theologians to guide individual (secondary) agents who needed to decide whether to participate or "cooperate" with other individual (primary) agents who were either actively involved in or contemplating doing something the church regarded as morally wrong. They are non-consequentialist, agent-relative principles, and in recent years have been adapted for use with collective agents as Catholic health care organizations began partnering with non-Catholic organizations. The principles require that a proportionate good be at stake (in this case, for example, the new knowledge to be gained by stem cell research for the health of future patients), or that there be some proportionate evil to be avoided (for example, the potential loss of at least some Catholic health care organizations if they cannot compete successfully with the non-Catholic organizations that offer human embryonic and fetal stem cell applications).[8]

Under the principles governing cooperation, the first and most important criterion concerns the intention of the Catholic secondary agent: the secondary agent should not intend the wrongdoing of the primary agent. This criterion reveals the non-consequentialist structure of these principles. In Catholic teaching, one should never intend an evil or intentionally do a wrong, even to bring about a great good or to avoid a great evil. Intending the wrongdoing of the primary agent is considered formal cooperation, and it is always impermissible. However, when the secondary agent intends a proportionate good or intends to avoid a proportionate evil by cooperating with the primary agent, cooperation is said to be material and thus may be permissible if additional criteria are satisfied.

There is no need to review all the principles and criteria here.[9] However, because the second principle has been at the center of some controversy in the past several years, we should briefly examine how it is used to determine the licitness of cooperation in wrongdoing. As the principles governing cooperation were adapted for use by collective agents, the emphasis on the agent's perspective has weakened somewhat, and more emphasis has been placed on using the principles to justify the agent's cooperation to outside observers, namely, the bishops (ordinaries) who oversee and are ultimately accountable to the Vatican for such choices by Catholic health care organizations. Under the second principle, cooperation by the secondary agent must be judged to be mediate rather than immediate. This distinction is intended to help us assess the degree or extent of participation by the secondary agent in the primary agent's wrongdoing. In practice, if an outside observer can distinguish the secondary agent's action from the primary agent's action, then the secondary agent's action is said to be mediate and may be permissible; if not, then it is said to be immediate. In 2001, after much internal debate and with pressure from the Vatican, the American bishops decided that immediate material cooperation in wrongdoing would henceforth be unacceptable for Catholic health care organizations (it is still part of the Catholic moral tradition, however, and as such is available for use by individual Catholic agents and other types of Catholic organizations).

With this brief background on cooperation, we return to O'Rourke's article. He suggests—more with a series of rhetorical questions than with a direct argument—that the use of human fetal stem cells could be licit as a form of mediate material cooperation, provided the stem cells were procured from cell lines and not immediately or directly procured from aborted fetuses. As the above criteria suggest, at stake here is not only the intention of the Catholic secondary agent but also how an outside observer might judge that intention if the secondary agent's actions are too closely associated with the primary agent's actions (O'Rourke uses the terms integral and accidental involvement to distinguish immediate from mediate material cooperation).

O'Rourke's conclusion on the question of the use of human embryonic stem cell lines is even more guarded. He refers to a document from the Pontifical Academy of Life, which determines that the use of human embryonic stem cell lines is not morally licit for Catholics. And, while he allows that this document is "not entirely helpful" and that it does not represent official church teaching on the matter, he says it should be considered carefully. Perhaps because of this church document, he does not follow the logic of his argument to its end. Again, he merely hints, though never clearly argues, that using human embryonic stem cell lines might also constitute mediate material cooperation and thus could be licit for Catholic health care organizations.

It is not necessary for my purposes that O'Rourke explicitly argue his case, because his implication seems clear. Here, I'm more interested in what he does not consider, namely, whether cooperation is the correct category of analysis. He merely assumes it is. I return to this question below, after we consider the argument of Professors Vincent Branick and M. Therese Lysaught.[10] They use the category of complicity to urge a conclusion that is opposed to O'Rourke's. They argue that Catholic health care organizations should not licitly use human embryonic stem cell lines, at least at this time (I will also return to this important quali- fication below), because their use, these authors claim, would make Catholic agents complicit in the original derivation pro- cesses that destroyed the embryo.

Branick and Lysaught begin by introducing the notion of complicity for their health care readers, noting correctly that it is not a category typically used in Catholic health care circles. It was invoked by the Pope and by some American bishops in their response to President Bush's compromise on federal funding for research on human embryonic stem cells,[11] and the president and CEO of the Catholic Health Association, the Rev. Michael D. Place, used the term when he affirmed the bishops' judgment in a statement to the press.[12] These invocations alone seem to have brought it into current usage in health care circles. In any case, the Pope, the American bishops, and Fr. Place seem to be using complicity in a normative sense. That is, they seem to be claim- ing that the use of human embryonic stem cell lines by Catholics would be wrong simply because it (somehow) involves the user in the derivation processes. They do not, however, provide us with the criteria that would help us understand the grounds on which they base their claim. Branick and Lysaught offer a more detailed argument. They too use complicity in a normative sense, although they provide criteria to help us assess relative degrees of complicity. What they do not provide is clear definition or explanation of complicity, and I will argue that this ultimately undermines the force of their argument.

Branick and Lysaught claim that complicity in Catholic teach- ing is "related to but not synonymous" with cooperation, but they do not tell us exactly how it is related, nor do they clearly define the circumstances under which it justifiably can, or should be, invoked. They do distinguish between formal and material com- plicity, and they suggest that these terms are used in ways analo- gous to their use in the principles governing cooperation (where, as we saw, they refer to the secondary agent's intention). However, as they go on, it seems clear that the terms formal and material are not, in fact, used with complicity as they were used with cooperation. Rather, formal complicity is said to refer to actions "before the fact" (one must presume) of the wrongdoing by the primary agent and thus to these authors implies an evil intention. But the presence or absence of a specific evil intention before an action that is later judged to be wrong is far from certain. I could be intentionally complicit in helping to plan a bank robbery, but perhaps not in the murder that takes place during it by one of my partners, when there was an agreement among the robbery team not to use violence. (I leave aside the question of whether I could be morally or legally accountable for the murder in any case—I am merely interested in showing that one cannot assume evil intentions are always present "before the fact.") Moreover, what Branick and Lysaught call material complicity is said to be complicity "after the fact," and presumably this weakens a claim of intention, though this too is far from certain. I could intentionally participate in the laundering of money from the same bank robbery, knowing that an innocent bank employee was murdered during the robbery. Both claims, I think, would need to be assessed on a case-by-case basis. Some examples or further explanation would have been helpful.

After this introduction, Branick and Lysaught list James Burtchaell's four types of complicity and discuss them briefly. They are: active collaboration in the wrongdoing; indirect association with the wrongdoing, implying approval; failure to prevent the wrongdoing when it is possible; and shielding the perpetrator from penalty.[13] They claim that the first two types are most relevant for the question considered here, where their so-called formal complicity would result from active participation in the derivation of human embryonic and fetal stem cells, and their so-called material complicity might result from using the stem cell lines derived by non-Catholics, if it could be argued that such use led to an indirect association implying approval.

Note, however, that for Branick and Lysaught material complicity is impermissible. Thus we might ask how "indirect" the association must be before they would judge material complicity to be sufficiently distant to permit Catholic health care organizations to use the stem cell lines. Branick and Lysaught offer no definitive answer to this question, though their criteria can be used to help us assess degrees of complicity—the more distant or remote, the less complicit is the Catholic agent's association. The criteria are listed as points to be considered and not in order of importance, and include the following: the time interval between the wrongdoing and the complicit actions; steps of separation between the wrongdoing and the complicit actions; whether the wrongdoing is a one-time action or an on-going practice; negative impact on the social fabric if one does not become complicit in some prior wrongdoing; the nature and immediacy of the goods or evils at stake;[14] and, the severity of the act.

Unfortunately, as I suggested above, their discussion of complicity is helpful but ultimately undermined by a certain lack of clarity about the category of complicity itself. Not only have they not given us a clear definition of what constitutes complicity in a wrongful action, they have not told us what makes complicity in such actions wrong, especially complicity "after the fact." It seems clear how being complicitous in the derivation processes for embryonic stem cells is wrong for Catholic health care organizations; they are not permitted to participate in or even be associated with the killing of human embryos. But it is not clear (to many) why it would be wrong for Catholic health care organizations to use the embryonic stem cell lines derived by other agents, or to use stem cell lines derived from aborted fetuses, when they had absolutely nothing to do with the derivation processes, the abortion, or the decisions related to them. Indeed, there is precedent for this. The church permitted Catholic health care organizations to provide and individual Catholics to accept a vaccine that was created with aborted fetal tissue.[15] The closest Branick and Lysaught come to an explanation of why complicity in the use of stem cells derived from human embryos or aborted fetuses would be wrong is that it would "lend legitimacy" to the original acts of killing the embryos or aborting the fetuses. This may or may not be true in fact, however, and whether one action would lend legitimacy to another is, in part, an empirical question. Would using human stem cell lines from prohibited sources be permissible if it could somehow be demonstrated that it did not lend legitimacy to the original acts? I suspect that Branick and Lysaught would resist this conclusion, and certainly many in the Catholic Church would.

    At this point in the discussion, then, we have Catholic authors using different categories of moral reasoning to arrive at conflicting, albeit guarded and preliminary, conclusions. In order to move the discussion forward, I submit that we need to decide whether cooperation or complicity is the relevant category of analysis[16] and how, if at all, cooperation might relate to complicity. And, we need greater clarification about the notion of complicity itself—admittedly, a difficult concept to understand and use—and about what makes a complicit action wrong. Professors M. Cathleen Kaveny and Christopher Kutz provide insight into these questions, and suggest a way forward to address the question considered here.

The Relation of Cooperation and Complicity

    Kaveny introduces a helpful new category of analysis to the Catholic tradition, which she refers to as the appropriation of evil, and grounds this category in an agent-relative, virtue-oriented moral theory.[17] She calls appropriation the "mirror image" of cooperation, and distinguishes them in the following way. The category of cooperation should be used, she claims, when facilitating or participating in another agent's immoral action, while appropriation should be used when one is using the fruits of another's immoral action. Cooperation is thus the relevant category of analysis when considering present, on-going, or future actions, whereas appropriation is relevant when considering past actions.[18] Kaveny's argument helps us understand why O'Rourke's analysis of Catholic cooperation in the use of human stem cell lines derived by others is not persuasive, even on its own terms.

How can Catholic agents meaningfully be said to participate in or be involved with the past evil actions of other, completely independent agents, when using the fruits of those actions for their own good purposes? O'Rourke compares the use of human embryonic stem cell lines by Catholic health care organizations to the use of the very limited amount of useful data derived from Nazi experiments during World War II.[19] But why should the use of such data be viewed as "cooperation" in any sense with the Nazi researchers who originally collected the data? No additional experiments are contemplated or even under consideration, and those that were done were completed many years in the past. It is simply not obvious that cooperation is the relevant category of analysis in such cases.

Kaveny also discusses the use of the Nazi data, but as a case of appropriation. This is, intuitively, much more persuasive. By using this data, present-day agents would not be cooperating with the Nazi's original evil actions, but surely they would be using the fruits of these actions for their own, presumably good, purposes. This immediately casts the use of the Nazi data in a more troubling moral light and suggests what it is that makes appropriation, in principle, wrong. If I am appropriating the fruits of another's evil actions to benefit myself or others, I have reason to pause and consider the effects, over time, such actions may have on my own character as a moral agent. I may also have reasons to pause and consider the effects such appropriation may have on a wider circle of relevantly affected agents. If we do not pause and reflect on such consequences, we may eventually become morally hardened or cavalier, and it is this effect that ultimately makes appropriation wrong—it may corrupt our moral characters and it may make it easier to do wrongful acts in the future.[20] However, if I were to use cooperation to analyze past actions, I might always conclude that we would be free to use the results of some previous immoral action, no matter how monstrous—simply because I cannot meaningfully be said to cooperate with primary agents whose actions are indeed past.

This said, we also need to ask how cooperation and appropriation relate, if at all, to complicity. Kaveny believes that cooperation and appropriation should both be considered types of or instances of complicity in wrongdoing.[21] This is also intuitively appealing, but we will need to confirm it with an exploration of the notion of complicity itself. For this we turn to Christopher Kutz's recent book on the subject.

Christopher Kutz's Notion of Complicity

Kutz's argument is too complex to summarize briefly, but parts of it can be isolated to help us understand his notion of complicity, when it can and should be invoked, and, to some extent, how to assess complicit actions in moral terms.[22] As suggested in the discussion above, complicity can be understood on both an individual and a collective level, though Kutz is clearly most concerned about our complicity as individuals in collective acts that may wrong or harm others. In any case, for Kutz, complicity concerns the relation of individual or collective agents to wrongs or harms mediated by other individual or collective agents. More precisely, it concerns the degree of moral accountability that one (secondary) individual agent or collectivity of individual agents has for the wrongful or harmful actions of another (primary) individual agent or collectivity of individual agents.

The awkward and repeated inclusion of "individual" is deliberate, for it is intended to capture Kutz's understanding of how collectives are made up of individuals who "act together" or "jointly" in particular ways and how, in Kutz's view, individuals are finally and individually accountable for the collective acts of the groups they join or in which they participate. But Kutz also distinguishes between two different types of groups, which he discusses as structured and unstructured collectives, and the moral complicity of individuals participating in these types of collectives is grounded somewhat differently. Governments, companies, and cartels are examples of structured collectives, whereas aggregated groups of relatively uncoordinated individuals who pollute the environment or who make millions of purchasing decisions in various markets around the world are examples of unstructured collectives.[23]

For structured collectives, Kutz claims that "all collective action is explicable in terms of the intentionality of [the] individuals..." who make up a given group.[24] Collective or "jointly intentional" action is defined by the content of the intentions with which individuals act when they act together. When individuals regard their own actions as contributing to a collective outcome, they may be said to share or to be acting with what Kutz calls a "participatory intention."[25] A participatory intention is understood as a threshold condition for the collective action of individuals, and thus Kutz deliberately uses a weak understanding of it. The content of individuals' intentions does not have to be identical for each individual; there must merely be enough "overlap" in the content of the individual intentions for the action to be considered collective. Moreover, the individuals contributing to the collective goal need not believe that the goal is realistically attainable or that the goal is even worthy of their efforts in order to do their parts in advancing it. Kutz claims that:

> A set of individuals jointly G when the members of that set intentionally contribute to G's occurrence by their particular parts, and their conceptions of G sufficiently and actually overlap. Put negatively, a set of individuals can jointly intentionally G even though some, and perhaps all, do not intend that G be realized, or do not even intend to contribute to G, but only know their actions are likely to contribute to its occurrence.[26]

Given this minimalist view of joint or collective action, Kutz develops a theory of moral complicity, that is, a theory about how to ground individual moral accountability for participation in the acts of other individual or collective agents. To do this, he first distinguishes between responsibility and accountability, and then develops a relational understanding of moral accountability that allows for plural perspectives. Relational accountability is one form of moral responsibility.

Kutz argues that the notion of responsibility has both non-relational and relational senses. In its nonrelational sense, responsibility "refers to a set of internal psychological competencies a person must have in order to be answerable for [a] harm [or a wrong]." The second sense is relational, and it "refers to a set of normative, external affiliations, the duties of the agent to other surrounding agents."[27] It is responsibility in the second sense, the relational sense, to which Kutz restricts his notion of accountability. It permits him to take multiple or plural perspectives into account when he evaluates an agent's complicit action in moral terms. He argues that we should evaluate an agent's actions from the perspectives of various respondents. The respondents might be onlookers or observers, victims, or, interestingly, the agent himself or herself when reflecting on a given action. Kutz claims that the moral valence of any given action will have as much, or more, to do with the relationships of the agent, onlooker, and victim as, say, the nonrelational qualities of the agent's will (intentions, desires) or the consequences of the agent's action. "Responses are warranted to what we do and who we are, not because of some deep metaphysics of causal responsibility, but because of what our actions and gestures of repair indicate about the view we take of our relations with others."[28]

Kutz's understanding of relational accountability is important in grounding individual accountability for alleged acts of complicity, especially where an individual participates in a wrong that seems to cause no harm or where an individual's participation in a group's collective action cannot meaningfully be said to be directly or causally responsible for the harm caused by the group. As an example, Kutz considers at some length the case of individual accountability for the deliberate bombing of civilians in Dresden during World War II, after an Allied victory was assured.[29] He allows that any one individual flyer, especially those who came late in the raid (after the firestorm was already raging), could plausibly claim that his participation in the harm the bombs caused was immaterial to the harm caused by the collective effects of all the bombs taken together. Such an individual, Kutz claims, made an "imperceptible marginal difference" to the harms actually caused, and if viewed from the agent's perspective alone might lead one to conclude that individual flyers could not be accountable for the harm caused by their collective actions.[30] Nevertheless, Kutz argues that they are morally accountable as individuals. He bases this on two arguments. First, the flyers individually intended the harm that their collective action achieved, at least in the weak sense described above but also, in some cases, in a much stronger sense. Second, he considers relationally both the perspectives of the victims, most of whom were innocent noncombatants, and the perspectives of the onlookers, represented by contemporaries of the event who knew the bombing was not required to win the war and by those of us today who enjoy the luxury of historical distance from the event.

Kutz also considers an example that is perhaps a bit more common place, but nonetheless compelling. He considers the case of a midlevel engineer who works for a company that makes electronic control modules. The engineer designs these modules, which his company uses to make consumer products, but which are also used to manufacture land mines that are sold in third-world countries.

> The relation between the engineer and the harm [caused by the land mines] is indirect, for he does not promote the mine sales intentionally if he does not know about them. However, there is still a significant basis for his accountability

for the mine sales, grounded in his intentional participation in a collective project....Indeed, so long as the decision to work with the company is voluntary and information about the company's activities is available, every employee bears an accountable relation to the victims of the land mines.[31]

In a relational understanding of accountability, Kutz claims, there is "[n]o participation without implication."[32]

Unstructured collective acts are similar to structured collective acts in that the individuals comprising the collective may not make a causal difference to the harm or wrong caused by it. The key difference is that individuals participating in unstructured collective acts may not be intentional participants, even in Kutz's weak or minimal sense. As an example, he appeals to a case in which individuals are polluting the environment by driving fossil-fuel burning vehicles, no one of which is causally responsible for the harms caused by the use of millions of such vehicles.[33] For individuals acting in this kind of collectivity, individual accountability for the harms or wrongs caused by the collective is established in two ways. First, Kutz expands the notion of individual participation to include participation in a culture or way of life. Second, he appeals to the individual agent's character. If I read him correctly, Kutz is making a virtue-based argument. The general intuition is that individuals are accountable for the social networks in which they are embedded, in the sense that these social networks are themselves products of human choice. Then, once individuals are made aware of how their participation in such networks both benefits them and causes harm to others, their moral characters can be subject to their own critiques and to those of other respondents, such as victims and onlookers from our society or from other societies. Such choices may reflect negatively on the characters of the individuals who made them, as viewed from the plural perspectives discussed above, and this may be enough to provide these individuals with the motivation to avoid such participation or, if this is not possible, at least to work to minimize or correct such harmful collective actions.

Assessing Complicity in the Use
of Embryonic and Fetal Stem Cells

Kaveny and Kutz give us ways, then, to clarify and extend the preliminary arguments of O'Rourke, Branick, and Lysaught. Together, Kaveny and Kutz provide us with a clearer understanding of complicity, when it can be invoked, what makes acts of complicity morally problematic, and how we can be held morally accountable for acts of complicity. Kaveny helps us understand what she calls the "mirror" of cooperation by introducing the notion of appropriation, and it now seems clear how the notion of complicity can include both cases of cooperation and appropriation. Both cooperation and appropriation are concerned with the participation of secondary agents in wrongs or harms mediated by primary agents, with appropriation being properly applied to the past wrongful or harmful actions of primary agents from which secondary agents benefit, and cooperation being properly applied to present or future wrongful or harmful actions that secondary agents might support for their own good reasons. Further, in linking appropriation to virtue theory, Kaveny helps us understand why complicity is wrong—because of its negative effects on the moral characters of agents. Kutz develops a similar argument about moral character when he discusses unstructured collectivities.

Assessing claims of complicity or degrees of complicity has not been discussed at any length, but we have the criteria under the principles governing cooperation and appropriation, and the points to be considered listed by Branick and Lysaught to which we can appeal. Kaveny also provides a list of questions to consider in assessing cases of appropriation. Kutz's discussion was directed mainly at establishing the grounds for moral accountability in complicit actions, but his understanding of relational accountability provides a place to begin the assessment process. Interestingly, adopting his understanding of relational accountability implies that the assessment process may itself need to be relational. In any case, given these resources and clarifications, we turn briefly to the question posed at the beginning of this article: Would a Catholic health care organization's use of the embryonic and fetal stem cell lines derived by other agents make them complicit in the derivation of these stem cells? Again, we may not be able to answer such questions definitively—not because we cannot reason to a conclusion, but because the notion of relational accountability may require extended dialogue among the persons or groups representing Kutz's various perspectives. Nevertheless, I may be able to sketch a response to it.

All of these authors emphasize the importance of intention in assessing an allegation of complicity, and Kutz is helpful in showing us why this is important in cases of structured collectives. The intention of secondary agents in the wrongful or harmful actions of a structured collective primary agent provides grounds for a judgment of complicity in the actions of the collective. For the question under consideration here, this alone should lead conscientious Catholic agents (individual and collective) to avoid business ventures with agents who derive embryonic or fetal stem cells, as the business venture would constitute a structured collectivity. Assuming the validity of Kutz's analysis, there is no way that observers of such relationships could avoid the conclusion that Catholic secondary agents are intentionally participating in the derivation processes of their business partner. This conclusion is consistent with the two preliminary appraisals reviewed above and will probably not be controversial in most Catholic health care circles.

However, the question of whether it is permissible for Catholic agents to use the embryonic and fetal stem cell lines derived by others is more difficult to assess. The two preliminary appraisals discussed above arrived at different conclusions on this question. But consider the following possibility. Presumably, Catholic agents would procure the stem cell lines in question from suppliers, and this suggests that such actions would constitute participation in a market. Such a market is an example of Kutz's unstructured collective. Participation, then, in such a market could potentially be accomplished without observers necessarily concluding that the Catholic agents were intending the wrongs or harms of the primary agent's derivation processes. Nevertheless, if we follow Kutz's reasoning, such participation would still constitute a form of complicity. In Kaveny's terms, it could probably be categorized as appropriation of the fruits of immoral derivation processes. Yet, one can ask an additional question, one that Kutz only hints at from time to time. In cases of participation in unstructured collectives, we may conclude not only that a secondary agent's participation in a primary agent's wrong or harm constitutes complicity, but also that the complicit action is so remote as to lead us to question whether the secondary agent can meaningfully be held morally accountable for it. In Catholic terms, such an observation could lead us to distinguish between what might be called licit and illicit complicity—in other words, it might lead us to grant the claim of complicity in the wrong or harm of the primary agent but not to conclude that it is therefore morally impermissible. If this suggestion is plausible, could it help us with the case in question?

I think it is possible that it could, providing certain additional conditions were satisfied (I am assuming for this assessment that the prior question of whether the stem cell lines in question are proportionate goods is not at issue). First, we would need to be satisfied empirically that the Catholic use of embryonic and fetal stems cell lines was not contributing to the death of additional embryos or fetuses, that is, more deaths than would have been the case without Catholic participation. It may be impossible to assess this question in any rigorous way at this point in time, although the following observations may prove relevant. With current federal regulations, we have reasonable assurances that the decision to abort is and will remain unrelated to the decision to donate the remains of a fetus for research or therapeutic purposes (unrelated, that is, from the perspectives of those procuring those remains). Also, so long as embryos are not being deliberately created or cloned for the stem cells that can be derived from them, the embryos from which the stem cell lines are derived were not going to be given the chance to develop in any case; in other words, these are embryos that will be discarded or frozen indefinitely. I do not wish to seem morally cavalier in these observations, or minimize the tragedy and wrongfulness of such deaths from a Catholic perspective. I merely wish to observe that the Catholic use of the stem cell lines derived from such sources is not likely to contribute to more embryonic or more fetal deaths than would have been the case without their use. Of course, this implies that should practices in this country change significantly, this condition may be impossible to satisfy. That is, if embryos begin to be created or cloned in large numbers expressly for their stem cells, or if fetuses begin to be deliberately conceived and aborted for theirs, Catholic health care organizations could not justify their procurement of such cell lines, for doing so could contribute to additional embryonic and fetal deaths. Complicity in such actions would not be permitted by the Catholic Church.

Second, Catholic participation in such markets should end immediately if other, less morally objectionable means are found to reach the same research and therapeutic goals that, it now appears, will be reached only through the use of embryonic and fetal stem cells.[34] This condition may not be acceptable to some Catholics because it may be construed in consequentialist terms. That is, it seems to suggest that the research and therapeutic goals of such research may be beneficial enough to justify the use of the stem cells derived from these problematic sources. However, it need not be viewed in this way if we are convinced that the first condition is satisfied. If the first condition is satisfied, then the second condition amounts to a recognition of what might be called "reluctant" or perhaps even "tragic" complicity in actions the Catholic agent would rather not be involved in if there were any other way of reaching the same goals. It is also a step toward addressing the concerns that are the focus of the third condition.

The third condition tries to address the observation made by a number of the authors discussed here, namely, that complicity in wrongful or harmful actions is wrong at least partly because it can undermine the moral character of the secondary agents. Thus, should Catholic agents decide to use embryonic and fetal stem cell lines, the third condition states that they should take steps to ensure that they and their subordinate agents do not become morally corrupted by such complicity. How might this be done?

Conscientious agents who reluctantly become complicit in the wrongful or harmful actions of others for some good purpose of their own can take a number of steps to lessen the chance that, over time, their moral characters will be adversely affected. Catholic health care organizations can address this by the care they demonstrate in arriving at and explaining their decision to use embryonic and fetal stem cell lines. Indeed, such a decision should not be carried out in secret, but should be communicated openly and publicly. The first and second conditions above could be emphasized in communications to and education for those using the stem cell lines. An explanation should also be offered to the bishops and sponsors who oversee Catholic health care, as well as to lay Catholics and others working in or observing Catholic health care organizations who might have moral concerns about such use. These actions might also address a common Catholic concern that such participation, even if justified, might lead to scandal.

Scandal is a technical term in the Catholic Church, and it refers to those attitudes or actions taken by some Catholics that lead others to do evil.[35] There are at least two possible responses to scandal. One is to avoid the objectionable action altogether and the second is to educate those who might be scandalized in ways that helpfully inform their consciences. The educational efforts undertaken to address condition three could also work to reduce the risk of scandal. This observation brings us back to the claim mentioned above by Branick and Lysaught, namely, that time can be a morally relevant factor in assessing claims of complicity. Time as such may not make much difference in these assessments, but efforts undertaken or changes that occur over time might. Education is one example of these efforts. Kutz's development of a relational notion of accountability suggests another. Relevant respondents can engage in dialogue about the meaning and assessment of complicit actions.

Claims of complicity, especially claims of the type we are considering in this case (that is, with unstructured collectivities), can be very difficult to assess morally. However, Kutz's notion of plural perspectives suggests that dialogue among those who represent the various perspectives can make a difference in the final assessment of complicit actions. Consider again the use of the data derived from Nazi experiments on innocent victims, data that some present-day researchers propose to use for good purposes.

Initially, a set of respondents might be horrified at the thought of using this data, especially if the respondents are the surviving victims of Nazi concentration camps. But suppose such victims meet the researchers who are proposing to use the data, and they find them not to be cold and calculating, intent only on new knowledge, but sensitive, humane researchers who are only too aware of the pain they might be inflicting on these victims. Because of this awareness, suppose also that the researchers decide to seek the guidance of the victims and their families. Furthermore, suppose that they explain privately to the victims and publicly to outside observers how and why they will use the data, and in so doing demonstrate great sensitivity toward the perspectives and moral assessments of the victims. One can imagine that slowly, perhaps even reluctantly, the victims might reverse their objections and eventually support the use of the data. Such a case might still be assessed as one of complicity in the original experiments, but one in which the researchers are not held morally accountable for their complicit actions by the victims or other observers of the process.

Dialogue among relevant respondents could, over time, change an assessment of a complicit action and, at the same time, it could be instrumental in helping to preserve the moral characters of the complicit agents. It is conceivable that a similar process might address the objections of Catholic observers of the Catholic health care organizations that use embryonic and fetal stem cell lines. The Catholic agents who actually use the stem cell lines for research or therapy will need to exercise the same kind of care and concern for the assessments of other respondents as the researchers did in the hypothetical case just considered. In doing so, they might preserve their own moral characters and, at the same time, persuade Catholic observers that their complicit actions ought to be assessed in such a way as to permit the agents to use, however reluctantly, the embryonic and fetal stem cell lines, subject to the above conditions. This could become a case of what, from a Catholic perspective, we might call licit complicity—a case where the actions of the Catholic agent are clearly complicit in some wrong or harm, but nevertheless are assessed on the basis of a set of criteria to be permissible.

Conclusion

If Catholic health care organizations use the embryonic and fetal stem cell lines derived by others they cannot avoid being morally complicit in the processes by which the stem cells were derived. But would they be complicit to the extent that their complicity should preclude their use of these stem cells altogether? Ultimately, if we follow Kutz's argument, this is a question that can be answered only in dialogue with other relevant respondents, though I have suggested some conditions above that, if addressed, might contribute to that dialogue and might permit Catholic agents licitly, if reluctantly, to use embryonic and fetal stem cell lines.

Notes and References

[1]Human stem cells derived from other sources are not problematic in the same ways, of course, and are not considered here.

[2]Catholic health care organizations are prohibited from using IVF technologies to create human embryos artificially, from destroying human embryos created either naturally or artificially, and from using the tissue of aborted fetuses. See United States Conference of Catholic Bishops. (2001) *Ethical and Religious Directives for Catholic Health Care Services*, Fourth Edition, Directives 38-41, 45, and 66. United States Conference of Catholic Bishops, Inc., Washington, DC.

[3]For a summary of the processes of derivation, see the National Institutes of Health, Department of Health and Human Services, *Stem Cells: Scientific Progress and Future Research Directions*, June 2001.

[4]This is not merely a self-interested concern, although it could be viewed as such. Catholic sponsors believe their organizations are continuing the healing mission of Jesus, and the loss of the health care ministry or some significant portion of it would be viewed as the loss of a significant moral and religious good.

[5]Whether it is morally acceptable to use human stem cells from a cultured stem cell line, as defined and permitted for federal funding by President Bush, is hotly debated among Catholic ethicists.... The Church's magisterium has not as yet specifically addressed this question." Albert S. Moraczewski, "Human Cloning and Stem Cell Research," supplement to Cataldo, P. J. and Moraczewski, A. S. eds. (2001) *Catholic Health Care Ethics: A Manual for Ethics Committees*. The National Catholic Bioethics Center, Boston, p. 7/4.

[6]It is possible to craft an alternative argument regarding the derivation of human embryonic stem cells by trying to persuade Catholic bishops that the church's position on the status of human embryos is not defensible, but it is highly unlikely that this argument could succeed.

[7]O'Rourke, K. D. (2001) Genetics and ethics, *Health Progress* **March-April,** 28–32.

[8]The proportionate consideration above introduces a consequentialist-like calculation into the principles, but the conclusion to the calculation tells the agent only that cooperation can be considered; it does not justify the cooperation as such.

[9]Readers might wish to explore these interesting and complex principles further in an article referenced in the discussion below: Kaveny, M. C. (2000) Appropriation of evil: cooperation's mirror image, *Theological Studies* **61**, 280–313. In developing her position of the appropriation of evil, Professor Kaveny helpfully summarizes and contrasts it to cooperation with evil.

[10]Branick, V. and Therese Lysaught, M. (1999) Stem cell research: Licit or complicit?" *Health Progress* 37–42. (This article was printed improperly, with some text missing and some repeated. It can be found intact at http://www.chausa.org/PUBS/PUBSART. ASP?ISSUE=HP9909&ARTICLE=B.)

[11]*See* "Stem-Cell Plan Termed 'Complicity in Evil,'" Catholic Health World (September 4, 2001). This article can also be found at http://www.chausa.org/PUB/PUBSART.ASP?ISSUE=W010901&ARTICLE=F.

[12]*See* "CHA Disappointed by Stem Cell Compromise," Catholic Health World (September 4, 2001). This article can also be found at http://www.chausa.org/PUB/PUBSART.ASP?ISSUE=W010901&ARTICLE=E.

[13]Burtchaell discusses complicity in relation to the use of aborted fetal tissue. *See* Burtchaell, J. T. (1989) The use of aborted fetal tissue in research and therapy, in *The Giving and Taking of Life: Essays Ethical*. University of Notre Dame Press, Nortre Dame, pp. 155–187.

[14]This criterion is not entirely clear. They give an example of a janitor who cleans hospitals where abortions are performed and claim the janitor would not be complicit in abortions that take place there.

[15]*See* Kenny, J. (1999) Accepting vaccination OK, according to ethicist. *St. Louis Review* December 24, 1999.

[16]Here, I ask this question from a Catholic perspective. Other options are available to non-Catholics and, interestingly, to Catholics not working in Catholic health care organizations.

[17]Kaveny, "Appropriation of Evil."

[18]One could appropriate the fruits of future actions, of course, but these

actions would still be in the past at the time the fruits were appropriated.

[19]O'Rourke, "Genetics and Ethics," p. 31.

[20]Kaveny explores some of these effects in an interesting discussion of how a secondary agent's character is morally "contaminated" by "seepage" and "self-deception," Kaveny, "Appropriation of Evil," pp. 305–306.

[21]Professor Kaveny related this to me in conversation.

[22]Kutz is more concerned with grounding individual complicity for collective harms or wrongs than he is in assessing that complicity in moral terms. He also interested in the legal application of complicity, but this topic, though interesting, is beyond the scope of this discussion. *See* Kutz, C. (2000) *Complicity: Ethics and Law for a Collective Age.* Cambridge University Press, Cambridge.

[23]Kutz, *Complicity*, p. 166.

[24]Kutz, *Complicity*, p. 71.

[25]Kutz, *Complicity*, p. 74.

[26]Kutz, *Complicity*, p. 103.

[27]Kutz, *Complicity*, p. 18.

[28]Kutz, *Complicity*, p. 139.

[29]Kutz, *Complicity*, pp. 117–120.

[30]Kutz is also concerned with his claim that any agent-centered morality, especially one that understands moral accountability merely in terms of those intentional wrongs or harms caused by agents, will not be adequate to account for our common moral intuitions about individual complicity in collective wrongs or harms. The multiple perspectives are part of his response to this concern. Kaveny's notion of appropriation is also concerned about moral accountability for actions not caused by the agents in question, though she remains committed to an agent-centered morality.

[31]Kutz, *Complicity*, pp. 156–157.

[32]Kutz, *Complicity*, p. 122.

[33]Kutz, *Complicity*, pp. 166–167.

[34]I credit this condition to Professor LeRoy Walters, who made this point at a meeting we attended together in Washington, DC.

[35]United States Conference of Catholic Bishops, *Ethical and Religious Directives for Catholic Health Care Services*, note 45, p. 43.

Abstract

Many of those who seek the regulation of human embryonic stem cell research express concern that human cells, tissues, embryos, and gametes are becoming commodified. To understand this concern, it is first necessary to consider both the role of market forces in human stem cell research and current regulation and legislation covering that research. This work analyzes the debate about commodification from a feminist bioethical perspective informed by Marx's theory of commodification and exploitation. Much of the discussion about commodification of human body parts, reproductive capacity, and reproductive tissues draws on the work of Anderson and Radin. I examine the capacity of those arguments to address the issue of commodification as it arises in human embryonic stem cell research. I then critically assess Dickenson's contribution to this debate, in which she argues that the current regulation of stem cell research is unjust toward women because it alienates them from their reproductive labor. I agree that one criticism of the commodification of human reproductive capacities is the alienation of labor. A more significant critique of the commodification of human reproductive tissues in stem cell research can be discerned through closer attention to the kinds of resource or product being commodified. Women's reproductive capacities are inevitably involved in generating the constituents of human life (ova, fetuses, and embryos). The lack of an appropriate means for recognizing the contribution of women's bodily capacities in the commodification of the resources they provide to stem cell research exemplifies an inherent contradiction of capitalist commodification.

Women, Commodification, and Embryonic Stem Cell Research

Susan Dodds

Introduction: Stem Cells, Demand
for Regulation, and the Market

One of the bioethical debates concerning stem cell research centers on the issue of commodification. In many calls for regulation of stem cell research, banning of human cloning and close monitoring of use of stored in vitro fertilization (IVF) embryos, there is a concern that humans, human life, or human tissues and capacities ought not to be treated as mere commodities that can be bought and sold in the marketplace. To do so, it is argued, is to fail to respect human personhood. In this way, the bioethical argument about stem cell research echoes some of the arguments connected with surrogate or contracted pregnancy, egg donation, and organ transplantation.[1] In this work, I examine these concerns about commodification and embryonic stem cell research from a feminist perspective, that is, I examine the social, politi-

From: *Biomedical Ethics Reviews: Stem Cell Research*
Edited by: J. M. Humber and R. F. Almeder © Humana Press Inc., Totowa, NJ

cal, and power relations involved in human embryonic stem (ES) cell research, with particular attention to the effects of such research on women and other oppressed groups.[2] Given the centrality of the concept of commodification to these debates, I go back to Marx's explication of commodification in capitalist societies to draw out the wider implications of the debate. I argue that there is within capitalism an inherent inconsistency between the understanding of each individual as a self-owning agent who is able to enter or refuse to enter contracts freely, and recognition of human subjectivity as embodied, material and socially interdependent. I draw on recent feminist writings on commodification to explore whether there are plausible routes out of that contradiction.

Women's Involvement in Development of Stem Cell Lines

As readers of this collection are aware, the scientifically fascinating feature of stem cells is that they can continually reproduce themselves. There are many kinds of stem cells in the body; some are more differentiated than others. "In other words when stem cells divide, some of the progeny mature into cells of a specific type (e.g., heart muscle, blood or brain cells), while others remain stem cells, ready to repair the everyday wear and tear undergone by our bodies."[3] ES cells are generally thought to be the most undifferentiated of stem cells, and they are often described a pluripotent.[4] ES cells, "unlike the more differentiated adult stem cells...retain the special ability to develop into nearly any cell type. *Embryonic germ (EG) cells*, which originate from the primordial reproductive cells of the developing fetus, have properties similar to ES cells."[5]

EG cells can be obtained from fetuses that have been aborted. ES cells can be obtained from embryos grown in vitro. These can be excess embryos left over from IVF treatment, human embryos created for the purpose of stem cell research or, potentially, either human or human/other animal hybrid embryos "generated asexually by somatic cell nuclear transfer or similar cloning techniques in which the nucleus of an adult human cell is introduced into an enucleated human or animal ovum."[6] To put it crudely, this process is basically the same as the technique used to produce "Dolly" the cloned sheep but without the technical difficulties of gestation and with the possibility of using different species to produce the ovum from the species whose cells are developed. ES cells are obtained from an embryo in a process that prevents cell differentiation in 5- to 6-day-old embryos, thereby stopping the development of the embryo, while providing a source of stem cells. It is the prevention of the further development of the in vitro embryo as a human embryo that constitutes the destruction of that embryo in the development of stem cells for research.

How do women participate in the production of ova and fetuses used in stem and germ cell research? ES cell research involves producing cells from embryos developed in vitro. The woman's participation in the research is similar to the process of IVF treatment (except that no embryo is transferred to her uterus). The collection of ova from the woman involves her taking drugs to induce hyperstimulation of her ovaries to "ripen" a number of ova, having her temperature monitored, and then undergoing either the surgical procedure of a laparoscopy or vaginal insertion of a long needle to enable ripe ova to be suctioned from her fallopian tubes.[7] In many cases, the embryos used are the excess embryos left over after a woman has achieved as many pregnancies as she wishes through IVF treatment or after she decides to abandon further IVF treatments. In EG cell research, the primordial germ cells (the cells that would have developed into gametes) of aborted fetuses are collected from the expelled fetus. In this case, the woman's participation involves the range of bodily, social, and psychological changes involved in early pregnancy, the decision to abort, the surgical process of abortion, and (at least where the regulations require researchers to seek consent to the use of aborted fetuses) the decision to make the fetus available for research.

Regulation of Stem Cell Research

The United Kingdom, the United States, Canada, and Australia (among English-speaking states) have all developed regulatory systems regarding research involving human embryos and cloning.[8] In the United Kingdom, research involving human embryos is regulated by the Human Fertilisation and Embryology Act (1990). In the United States, state laws determine the extent to which human embryonic stem cell research is permitted, but access to federal funds for research have been limited through the (now-defunct) National Bioethics Advisory Committee's 1999 recommendations and the Bush Administration's restrictions on embryo use.[9] In Canada, C-13, *An Act respecting assisted human reproduction*, and the Canadian Institutes of Health Research Guidelines, *Human Pluripotent Stem Cell Research: Guidelines for CIHR-Funded Research*, both introduced in 2002, frame research involving stem cells derived from human embryos. In Australia, federal legislation covering human embryonic stem cell research and cloning was adopted in 2002, and state-level legislation complementing the federal legislation is expected to be debated in 2003, while ethical guidelines regarding research involving human embryos are under review.[10] To varying degrees each country's legislation or guidelines also regulate the use of fetal tissue in research and so affect researchers' access to EG cells.

In general, each jurisdiction has prohibited human cloning, but permits limited use of human ES and EG cells in the development of cell lines. In each of the guidelines and regulations listed above, if it is permissible for embryonic stem cells to be used in research, they are only to be obtained from excess IVF embryos. No embryos are to be created for the purpose of research, and women who consent to their surplus IVF embryos being used for stem cell research are not to be paid for providing the embryo or embryos to researchers. Women may, however, be paid for their time and inconvenience.[11]

These regulatory regimes shape how embryonic stem cell research is conducted, but they do not appear to impede research (except for research into cloning). In Australia, for example, the current guidelines prohibit the use of human embryos in the development of ES cell lines, cloning, and gene therapy on embryos. Therefore, for Australian scientists at the world-famous Monash Institute of Reproduction and Development to obtain embryos for their research, they had to import embryonic cell lines from Singapore.[12] There is no prohibition on conducting research on imported tissues or cell lines or on importing the tissues or cell lines. The new Australian legislation permits the development of some ES cell lines only where those stem cells are derived from surplus IVF embryos with the consent of the women or couple for whom the embryos had been originally created. The legislation is silent, however, on the importation of cell lines, whether such imported cell lines were created for the purpose of research, and whether the woman who provided the fetus or embryo from which the stem cell line was derived consented to such use or was paid for her contribution to the research effort.

Commodification of Stem Cell Lines

Although researchers, bioethicists, and policymakers alike agree that it would be a bad thing for human beings to be treated merely as property,[13] ownable by others, there are reasonable profits to be found in exploiting advances in biotechnology and medical research that use human embryos and fetal cells as a fundamental resource. Although the regulations prohibit the direct sale of fetal tissues and embryos through prohibiting women from entering the market to trade in fetal tissue and embryos they have produced, it does not prohibit trade in products derived from embryos or fetal tissue in the lab. Suzanne Holland refers to the emergence of markets in body parts, gametes, embryos, and cell lines that "on sell" donated material as "downstream commodification." Very often, she argues, the donors of body parts, embryos, gametes, and so on are not aware of the potential for their donation to enter the realm of commerce, nor are they aware of the profits to be made in these markets.[14] In the case of embryo donations that are used for research, Holland cites Nelkin and Andrews who write of the US situation:

> When IVF patients are asked to donate excess embryos for research, they are not told specifically what that research will entail. They generally assume that their embryos will be used by their own doctor to help other couples that suffer from infertility. They might be disturbed to learn that their embryos have been used to develop a commercial cell line.[15]

There is currently a market in ES cell lines. One source for the embryos from which those cell lines are derived is the United States, and another is Singapore. The recent Australian experience provides a glimpse into the market in stem cells.

In Australia, there are two key teams of human stem cell researchers. They are the Monash Institute of Reproduction and Development, already mentioned, and a team at the University of Adelaide. In both cases, researchers are working in partly privatized parts of Australian public universities, and receiving considerable support from Australian public research funds. The team of ES cell researchers based at the University of Adelaide is funded by BresaGen, a biotechnology company. The Sydney Morning Herald reported in July 2000 that the Adelaide researchers had paid $5000 (Australian dollars) for ES cells and in September 2000 that BresaGen had contracted with the University of Wisconsin for the supply of ES cells to the Adelaide researchers.[16]

Meanwhile the Monash Institute's efforts were attracting venture capital. The Monash Institute media release of August 11, 2000 announced that ES Cell International Pty Ltd (ESI) had formed a consortium of Australian and Singaporean investors who had "committed $17 million (Australian dollars) in seed capital to develop its research into ES cells being conducted" by the Monash Institute and its research partners in Israel and

Singapore. The media release goes on to explain that ES Cell Australia Pty Ltd is "an Australian private investor venture capital entity currently focussed on investment within the biotechnology environment." Its Singapore partner is Life Sciences Investment Pty Ltd (LSI), a "wholly owned subsidiary of PharmBio Growth Fund Pty Ltd, which is a joint investment fund by EDB Investments Pty Ltd and NSTB Holdings Pty Ltd. LSI has an initial fund of $20 million under management."[17]

> Its mission is to stimulate growth of the life sciences industry in Singapore through commercialization of technologies from local and foreign technolog[y] license[s], and to achieve medium to long-term capital gain. LSI aims to create long-term value by assisting entrepreneurs in building companies from the formative stage.[18]

In the United States, the Geron Corporation, which funded both the University of Wisconsin ES cell line research team and the Johns Hopkins EG cell line research team, is seeking to patent the technology whereby ES cell lines have been developed successfully.[19] A patent of the process would ensure profits so long as stem cell research is conducted, regardless of any regulation of markets in human tissues.

Clearly, despite widespread public, political, and scientific concern about the commodification of human tissues involved in ES cell research, there is a market and investment capital available for human embryos and fetal tissue products.

Commodification and Reproduction

Given this overview of current regulation and commercial interests in human ES cell research, there remains to be examined the matter of feminist accounts of commodification of human reproduction and reproductive tissues. This part of the work looks in detail at one of these approaches, with reference back to Marx's critique of capitalist production and the commodification of labor.

A number of feminists have written to express their concerns about the growing market in human bodily tissues in general and the market in human gametes, embryos, fetal tissues, and gestational capacities in particular. Many draw on the neo-Kantian writings of Margaret Jane Radin[20] and, to a lesser extent, Elizabeth Anderson's earlier work.[21] Anderson considers the debate surrounding contracted pregnancy, or surrogate motherhood, and argues against women's reproductive labor becoming a fungible market commodity, that is, that the value of women's reproductive labor ought not to be viewed as a value that is directly commensurable with some other value, such that a "fair price" or value can be set on such labor. Her view hinges on a Kantian conception of persons, their intrinsic value, and an ethic that rejects the use of persons as mere means.

> Contract pregnancy transforms what is specifically women's labor—the work of bringing forth children into the world—into a commodity. It does so by replacing parental norms that properly govern the practice of gestating children with the market norms that govern ordinary production processes. The application of commercial norms to women's reproductive labor reduces surrogate mothers from persons worthy of respect and consideration to dominated objects of mere use.[22]

There are two aspects to Anderson's argument: first, that surrogacy contracts interpose a market relationship into a relationship (that of mother and child) that should be governed by nonmarket values: parental love—"passionate, unconditional commitment to nurture one's child by providing her with the care, affection and guidance she needs to develop her capacities to maturity."[23] Second, that those who create and use the market in surrogacy fail to respect the women who might become surrogates as autonomous agents, worthy of respect. Women who enter restrictive, degrading contracts to carry children for others are treated as mere means to others' ends; their bodily and emotional autonomy is denied.

Given the emphasis on parental love and the restrictions on autonomy posed by surrogacy contracts in Anderson's argument, a question arises about its applicability to the commodification of human embryos and fetal tissue in stem cell research. If stem cells are derived from excess IVF embryos that the woman has chosen not to use in an attempt at pregnancy, then there is no parent–child or gestational relationship that is damaged by the market in embryos or stem cell lines. Similarly, assuming that women can legitimately and autonomously choose to terminate a pregnancy, there is no potential relationship with a child that is affected by the commodification of fetal tissues. Both those conclusions rest on two important premises, however. First, that human embryos or fetuses do not share the moral status of "persons" in the Kantian sense of autonomous beings that are ends in themselves. Second, that any commodified relationships concerning embryos or fetal tissues do not exploit women such that their autonomy is denied and they become mere means (embryo and fetus producers) to others' ends. Anderson's argument, when applied to the use of embryos and fetal tissue in human ES cell research, justifies limitations to the market in embryos and fetal tissues but not the prohibition of such commodification, even though her argument, as applied to contracted pregnancy, would exclude surrogate motherhood from the scope of the market.

Margaret Jane Radin also argues for setting limits to the scope of market relations, but her argument emphasizes the importance of regulation of some markets, rather than wholly excluding market relations from entire areas of activity. For Radin, the freedoms of market choice need to be tempered by consideration of other significant values that are not captured by market exchange. Those potential commodities which are recognized as having both a market value, and other, possibly incommensurable, values are to be understood as "contested commodities" and that their transformation into the market is incomplete, which justifies significant regulation.

A key difficulty in applying Radin's approach is that it seeks to occupy a middle ground between trade in fully fungible commodities and protection of certain types of things, practices or services from the logic of market exchange. Radin is right to state that a great many commodities that are exchanged in the market are not fully commodified. Purchase of land, many capital assets, or historical objects, for example, does not transfer to the owner unlimited rights over the goods—land must only be used in accordance with local planning, environmental and other public health conditions; capital assets are often subject to foreign ownership, anti-monopoly, or other restrictions; historical artifacts are often subject to preservation conditions. However, providing an ethical justification for the various conditions set on property rights, and hence the limits placed on the market, is a mammoth task, and it is not clear that Radin's argument has provided sufficient detail to support such a task.

Suzanne Holland seeks to use Radin's conception of "incomplete commodities" to argue for substantial governmental regulation of the market in body parts, specifically tissue, gametes, and embryos. She argues that existing government efforts at regulation in these areas are inconsistent, indirect, and inherently untenable.[24] She urges government to step into this role as she favors "government using its regulatory power to curb the private sector's interest in unregulated commodification of these contested commodities that are connected to our tenacious commitments to personhood."[25] As Radin comments in response to Holland, while the notion of incomplete commodification can be used to encourage policymakers to think about the issue of regulation of markets differently, in a nuanced way, that very approach can get in the way of providing a good justification for any specific regulation.

Yet with this ability to focus on nuanced regulation comes the characteristic pragmatic difficulty: the absence of any ability to use broad-brush general principles to determine in advance what ought to be done in any given case.[26]

In my view, a key limitation to the positions developed by Anderson and Radin is that they cannot provide a coherent justification for distinguishing the market in human tissues and reproductive capacities from the commodification of labor-power in capitalism. Although both are critical of the inequities created by the capitalist exploitation of labor, neither addresses the relationship between commodification of bodily capacities and bodily tissues in detail. Markets in body parts are treated as areas where either market relations pervert appropriate human relationships or as areas where market transactions ought to be carefully regulated so as to protect important human values. The market in human labor is not subject to the same scrutiny.

"Sweat Equity" and Women's Alienation From the Products of Their Labor

Donna Dickenson has recently provided an analysis of the commodity relations involved in ES cell research that does address the connection between labor and commodification of embryos,[27] but she does so, in my view, in a way that still misses a central conundrum of the conception of persons involved in these discussions of commodification of body parts. To get to that conundrum, it is worth examining Dickenson's argument in some detail.

Dickenson's argument hinges on the view that the "sweat equity" due to women for the labor of producing ova or embryos or fetal tissue is alienated from them in stem cell research. Dickenson's argument draws on Marx's conception of alienated labor and reproduction as well as Christine Delphy's (1984) development of a Marxist understanding of women's domestic labor as "the domestic mode of production." She then seeks to extend that understanding of labor to women's labor in providing the ova, embryos, or fetal tissues used in human ES cell research as a way of demonstrating women's entitlement to recognition and payment for their labors (as a matter of justice) rather than denying women an opportunity to enter the market. She argues that women are denied access to this recognition, and are open to exploitation, because property rights are not recognized in the body.[28]

Reproduction, Commodification, and Alienation

Marx distinguishes biological reproduction (reproduction of the species) from reproduction of the conditions of social production.[29] Under capitalist modes of social production, reproduction (in the second sense) involves reproduction of the labor force through social institutions that provide for the sustenance, education, and material needs of workers. Although this form of reproduction may include domestic or household labor, it is not reducible to it (e.g., workers could live in dormitories serviced by a paid cleaning and catering staff). Reproduction in this sense does not include biological reproduction.

To understand domestic labor as a site of women's alienation from their labor and their oppression, the Marxist account needs to be extended, as it is in Delphy's account,[30] by articulating a distinct mode of production, "the domestic mode of production," that can explain the value and role of women's domestic labor as a social institution. Delphy writes, "It is women as economic agents who are excluded from the (exchange) market, not what they produce."[31]

Although Delphy's argument identifies an inconsistency in conventional Marxist thought and offers a materialist argument for viewing unpaid domestic labor as alienated labor, it risks equating women with domestic work to the exclusion of other forms of work or production. Delphy challenges the view that women's work in domestic relations of (in Marx's sense) reproduction is natural rather than social. She does not challenge, however, the view that women will (naturally) be the people engaged in domestic relations of production, hence she does not challenge the heterosexism and gender essentialism presumed by the account, or the sexual division of labor.

Productive Reproductive Labor and Bioethics

Dickenson seeks to apply and extend Delphy's account of the alienation of women's labor in 'domestic relations of production' to issues in bioethics. For Dickenson it is the alienation of women's reproductive labor through policies that prevent women from exchanging their "sweat equity" in ova, embryos, or fetuses used in stem cell research that is unacceptable. In her view the focus on donation of reproductive tissues for this purpose contributes to women's poverty and oppression. That is, Dickenson argues, women are mistreated in stem cell and germ cell research because the labor of women that contributes to the production of embryos or fetuses is alienated from those women: women aren't paid for their labors. The emphasis on women's "donation" (where there is no avenue for payment), simply puts an acceptable face on the exploitation of women in the lucrative stem cell industry.

> A compulsory and one-way gift relationship is not gift, but exploitation. Nor is it adequate to conceptualize the issue in terms of consent alone, any more than it is adequate to say that the worker consents to work and therefore retains no further rights to control the conditions of his labour.[32]

It is worth discussing again what women's labor amounts to in stem cell research: taking super-ovulatory drugs, awaiting the ripening of one or more ova, being closely monitored, and undergoing a procedure for the removal of her ova. In many cases, the collection of ova will have occurred before the decision to make an embryo available for stem cell research, because the woman will have been receiving IVF treatment and the ova used for stem cell research have already been used to make embryos that are no longer required. In germ cell research, the primordial germ cells are collected from aborted fetuses. In this case, the woman's participation involves the range of bodily, social and psychological changes involved in early pregnancy, the decision to abort, the surgical process of abortion and (at least sometimes) the decision to make the fetus available for research.

Although some of what a woman does in these two processes is not a matter of volitional labor, but rather bodily processes (albeit tweaked by pharmaceuticals and surgery), Dickenson's commitment to breaking down the distinction between "what is natural to women" and "what is socially valuable labor" allows her to argue that women are alienated from the products of their labor. Dickenson argues that the regulatory prohibition on paying women for their ova or aborted fetuses perpetuates the false distinction between what women do naturally and productive labor. For example, the recent Australian legislation restricts stem cell research to use of embryos that are excess IVF embryos that would otherwise be discarded, and have been donated by the woman or couple. The legislation prohibits women from selling their embryos or aborted tissues and prohibits the creation of embryos outside of IVF therapy for the purpose of research. Similarly, the UK Polkinghorne Committee's recommendations seek to block the possibility of women gaining property rights, as Dickenson puts it, "in the products of their reproductive labor. Apparently, if that labor counts as labor at all, it lies outside the realm of productive work—into which the work of researchers and biotechnology companies comfortably fits."[33]

Dickenson does not argue for the full-scale commodification of reproductive tissues, opening up a free market in ova, embryos, and aborted fetuses. Rather she wishes to open up a wider range of options for women than the current single option (should they wish to contribute to stem or germ cell research) of donation. Her argument focuses on women's control over how the products of their reproductive labor are used. She canvases the following three options:

1. The status quo: reinforcing the law's primary concentration on obtaining a genuine informed consent from the tissue donor, rather than conferring property rights.
2. Stricter regulation of commercialization.
3. Vesting control over all tissue in the mother, and treating alienation from her as theft.[34]

The first two options are thought by Dickenson not to be adequate to her purposes. Option one is intended to give women greater control over how embryos are used and to ensure that women are fully informed about the nature and intent of the research being undertaken prior to relinquishing the tissues. In Dickenson's model this "continues to maintain what has become a fiction in actuality, if a fact in law: that tissue extracted after procedures is no longer of any interest to anyone" (2000, p. 213). The second option merely attempts to reform existing practice, but attempts to do so in the face of rising pressure for increasing the range of research conducted in this area. The third, although stopping short of recognizing women's property rights in reproductive tissues, is meant to strengthen women's power to determine how those tissues may be used, to transfer control over the tissues, and to exclude others from access to the tissues (generally, or specific others). Dickenson appears to be arguing that women's value-producing labor in these reproductive activities will only be recognized if a form of limited property in the tissues women produce is recognized. Until that recognition occurs, women will continue to be alienated from their labors.

I agree with Dickenson that one way in which women can be treated unjustly in stem and germ cell research is in the exploitation of their legislated lack of control over what happens to the reproductive tissues they produce. In that sense, women are alienated from their reproductive labor. I think, however, that a stronger argument can be made for recognition of women's contribution to these lucrative areas of research through the related Marxist concepts of exploitation and surplus value.[35]

A Return to Marxist Understandings of Commodities, Exploitation, and the Commodification of Labor

Gerald Cohen's work on self-ownership draws on the Marxist concept of exploitation and the view that the injustice of capitalist modes of production is centrally attributable to the expropriation of surplus value.[36] The capitalist, in effect, steals the labor time of the worker.

The argument has the following structure. Each person owns his or her own bodily capacities and labor-power. In wage labor, workers are paid for the use of their labor-power in production—applying their labor-power to natural or partly improved objects so as to appropriate the useful qualities of them. Those things can thereby become commodities and be exchanged. So, although labor-power creates the bulk of the exchange value of products, workers are not paid the full surplus value (that is, the change in the value of the commodity attributable to their labor). To put it another way, the wage a worker receives only covers a part of the increased value that results from their labor. The exploitation of workers and the injustice of capitalism is the expropriation of surplus value. When a section of society has no access to the means of production, such exploitation becomes the norm and capitalism flourishes.

Applying this understanding of the injustice of capitalist modes of production to women's reproductive labor and reproductive capacities in the commodified relations surrounding ES and EG cell research highlights the significance of at least one aspect of the ambivalence surrounding women's agency and capitalist commodification.

In volume one of *Capital*, Marx articulates both the nature of commodities and the commodification of labor under capitalism. Marx writes

> A commodity is, in the first place, an object outside us, a thing that by its properties satisfies human wants of some sort or another. The nature of such wants, whether they spring from the stomach or from fancy, makes no difference.[37]

This satisfaction of human wants is the use-value of a commodity. To be a commodity, however, a thing must also have exchange value: that is, it must be exchangeable for other things with use-value. A commodity is, finally, an object whose value is created through purposeful labor. For something to be fully commodified, it must be produced so that it can be exchanged on the market.

In late capitalism, the scope of commodifiable entities has expanded so that nearly every thing/process/idea can be exchanged in the market. Human tissues and biological capacities, including reproductive tissues and capacities, are not spared the spread of commodification (and it may well be that the pressure to commodify such tissues and capacities is particularly acute in developing nations where access to the means of production is most limited). Although regulations may exist to keep human embryos, ova, and fetal tissue out of the market, those tissues are used in the development of products, including cell lines and therapeutic agents that are recognized as commodified products.[38] The question, as Dickenson identifies it, is whether women who produce the ova, embryos, or fetuses used in the development of these commodities should be recognized as having property rights over the products of their reproductive labors. In my view, however, the discussion of commodification of ES and EG cells highlights two contradictions of capitalism. First, in recognizing a person's labor-power as something that is owned by the laborer as an agent and as something which the agent is forced to sell for their material survival. Second, by demonstrating the refusal to recognize the potential value of biological processes as meriting financial reward.

On the Commodification of Labor

In the chapter entitled "The Buying and Selling of Labor-Power" Marx writes:

> In order to be able to extract value from the consumption of a commodity, our friend, Moneybags, must be so lucky as to find, within the sphere of circulation, in the market, a commodity, whose use-value possesses the peculiar property of being a source of value. The possessor of money does find on the market such a special commodity in capacity for labor or labor-power.

> By labor-power or capacity for labor is to be understood the aggregate of those mental and physical capabilities existing in a human being, which he exercises whenever he produces use-value of any description.

> But in order that our owner of money may be able to find labor-power offered for sale as a commodity, various conditions must first be fulfilled. The exchange of commodities of itself implies no other relations of dependence than those which result from its own nature. On this assumption, labor-power can appear upon the market as a commodity, only if, and so far as, its possessor, the individual whose labor-power it is, offers it for sale, or sells it, as a commodity. In order that he may be able to do this, he must have at his disposal, must be the untrammelled owner of his capacity for labour, *i.e.*, of his person. He and the owner of money meet in the market, and deal with each other as on the basis of equal rights, with this difference alone, that one is the buyer, the other the seller; both, therefore, equal in the eyes of the law.[39]

The conditions required for the owners of labor-power to be able to exchange their labor in the market are that they retain their ownership of their labor-power so they can only sell it for a fixed period, not indefinitely, or they would be slaves, not owners. For the owner of money, Moneybags, the capitalist, to be able to find and use this commodity, "the laborer instead of being in a position to sell commodities in which his labor is incorporated, must be obliged to offer for sale as a commodity that very labor-power, which exists only in his living self."[40] As long as the laborer lacks the means of production, they will be forced to sell their labor-power. This opens up for the owner of capital the opportunity to exploit the worker, paying them less for their commodified labor-power than the value added, by their labor to the commodities produced. Thus, the laborer must own their own person, be an agent who is free to sell his labor-power and, for capitalist accumulation of wealth to occur, the laborer has to be forced to sell their labor-power because they lack access to the means of production.

As Marx puts it

> For the conversion of his money into capital therefore, the owner of money must meet in the market with the free laborer, free in the double sense, that as a free man he can dispose of his labor-power as his own commodity, and that on the other hand he has no other commodity for sale, is short of everything necessary for the realisation of his labor-power.[41]

Conclusions: Reproductive Tissues: Products of Labor or Natural Resources?

Where does this leave us in the stem cell debate? It appears that human stem cells could be thought of as a commodity with the peculiar property of having myriad use-values, as well as exchange value. However, the labor and resources used to develop these commodities do not fit into capitalist modes of production at all well. I believe we are left with limited options, none of which wholly avoids the conundrum that women are viewed as objects and subjects of property.

On the one hand, people are not commodities—rather, they are agents who can commodify their labor-power. However, women have been identified with their reproductive capacities, and reproduction—unlike labor—is not viewed as something which women choose to do or act within, despite Dickenson's attempts to change this understanding.[42] There are two alternatives here: human embryos and fetal tissue could be understood as products of women's bodily labor (to be fair, men need to be recognised as having made some contribution of labor, too). Alternatively, human embryos and fetal tissues could be thought of as a natural resource, produced by women (again with a bit of help from the natural resources produced by men).

On the exploitation approach it seems clear that, if women are viewed as agents who choose to labor so as to produce the tissues used in the development of cell lines, they are being exploited by the theft of the surplus value attributable to their labor. The requirement in current legislation that women's contributions be in the form of gifts and the policy viewing reproduction as a wholly natural process (as not in any way under rational control) plays on patriarchal attitudes toward women and reproduction and serves to preserve the profits of those who exploit this resource.

When compared with the surplus value created by the labor of the lab technicians and scientists involved, however, this exploitation is likely to seem negligible. Unless the technical skills and efforts of the scientists are used to transform the embryo or fetal gametes into a stem cell line, the embryo or fetal tissue is pretty worthless.

If women are not entitled to a say and a financial interest in the products of their reproductive labor then they could be viewed as producing a natural resource, which under capitalist terms has no value until it has been labored on by the technicians employed by the commercial interest so as to produce something of exchange value. But this appears to ignore the fact that there would be no human stem cell research without human ova, sperm, embryos, or fetal gametes. Furthermore, for the moment at least, that the only source of ova, embryos, and fetuses is women and men. But women are agents, not natural resources or mines.

The problem with being left with these two alternatives is that there is no possibility of commodifying stem cells and cell lines without women's contribution of their ova, embryos, or fetal tissue. If women are, like laborers, untrammelled owners of their persons, then they should not have their ownership trammelled by expropriation of their bodily tissues. The human body is a strange thing, however, it is both essential for our material existence and it can also be a source of resources and commodities. The range of regulations and legislation debarring women from profiting from their biological capacities attempts to draw a veil over this ambiguity. In doing so, however, they simultaneously protect women from being understood as analogous to mines from which resources can be extracted, and undermine women's claims to full agency and self-ownership. Unlike Dickenson, I do not see this as a problem unique to women and their reproductive capacities. The debate about commodification of reproductive tissues brings into stark relief one key contradiction of capitalism: that it relies on a conception of the worker as someone who is alienable from his or her labor-power, who must be understood both as an embodied agent, so that material necessity will drive the laborer to sell labor-power, and at the same time as a person whose bodily capacities are alienable in wage labor. Finally, I see it as identifying a contradiction in those strains of liberalism that conceive of persons as choosers who are conceptually prior to their social embodiment.

Is there a third option? Perhaps by going back to Marx on commodification there is scope for developing a view of women as artisans who also possess, by nature, the capital and resources required for the production of their goods. A woman who offers up her ova or fetal tissue for the purpose of stem cell research is an artisan whose labor is incorporated into the products she creates. Certainly a bit of sperm obtained from a man is required, but women are still entitled to a substantial portion of the exchange value of stem cell products because, without them, there would be nothing to exploit.

Notes and References

[1]*See*, for example, Nelkin, D. and Andrews, L. (1998) Homo economicus: commercialization of body tissue in the age of bio-technology, *Hastings Center Report* **28**, 30–39; Mackim, R. (1996) What is wrong with Dodds, women commodification and embryonic stem-cell research commodification?, in *New Ways of Making Babies: The Case of Egg Donation*, (Cohen, C. B., ed.), Indiana University Press, Bloomington, Indiana; and Holland, S. (2001) Contested commodities at both ends of life: buying and selling gametes, embryos and body tissues, *Kennedy Inst. Ethics J.*, **11(3)**, 263–284.

[2]Sherwin, S. (1996) Feminism and bioethics, in *Feminism and Bioethics*, (Wolf, S. M., ed.), Oxford University Press, New York, pp. 47–66.

[3]National Bioethics Advisory Committee (NBAC) (1999) *Ethical issues in human stem cell research: executive summary*, National Bioethics Advisory Council, Rockville Maryland, p. 1.

[4]There is debate about whether adult stem cells might be just as versatile or pluripotent as embryonic stem cells. *See* Dewitt, N. and Knight, J. (2002) Biologists question adult stem-cell versatility, *Nature* **416**, 354.

[5]National Bioethics Advisory Committee (1999), ibid.

[6]National Bioethics Advisory Committee (1999), ibid, p. 2.

[7]*See* Fischel, S. and Jackson, P. (1989) Follicular stimulation and high tech pregnancies: are we playing it safe? *BMJ* **299**, 309–300 cited in Resnik, D. B. (2001) Regulating the market for human eggs, *Bioethics* **15(1)**, 1–25; 3.

[8]National Bioethics Advisory Committee (NBAC) (1999) Ethical issues in human stem cell research, National Bioethics Advisory Council, Rockville, Maryland; Bush, G. W. (2001) Speech from Crawford, Texas, *Bush announces position on stem cell research*, Washington Post, August 9, 2001. Available online at www.washington post.com/wp-srv/onpolitics/transcripts/bushtext_ 080901.htm

[9]Canadian Institutes of Health Research (2002) *Human pluripotent*

stem cell research guidelines, Government of Canada, Ottawa, available at http://www.c ihr-irsc.gc.ca/publications/ethics/ stem_cell/stem_cell_guidelines_.shtml.

[10]Australia (2002) *Research involving embryos and prohibition of human cloning Act*; National Health and Medical Research (1996) *Ethical guidelines on assisted reproductive technology* Commonwealth of Australia, Canberra (these guidelines are currently under review in light of the *Research involving embryos and prohibition of human cloning Act*).

[11]Holland (2001) *op cit.,* and Resnik (2001) *op. cit.* offer interesting insights into the amount of money that can be offered in "compensation" for pain and use of a woman's time in the case of egg donation in the US, where it appears that some women can be paid many thousands of dollars for the eggs produced after one cycle of hyperstimulation.

[12]Fannin, P. (2000) Hope on diseases after lab triumph, *The Age,* August 15, 2000, Melbourne, Australia.

[13]In Kantian terms, treated as "mere means".

[14]Holland (2001) *op cit.* pp. 266–271.

[15]Andrews, L. and Nelkin, D. (2001) *Body Bazaar: The Market for Human Tissue in the Biotechnology Age,* Crown Publishers, New York, p. 35; cited in Holland, *p. cit.,* p. 269.

[16]Smith, D. (2000) Embryo stem cells bought from US, *The Sydney Morning Herald,* July, 31 2000, Sydney, Australia.

[17]Monash Institute of Reproduction and Development, (2000) International Joint Venture to Develop Cell Biology Breakthrough, Media Release, August 11, 2000, available at http://www.monash institute.org/media_releases.cfm?doc_id=81.

[18]Monash Institute of Reproduction and Development, (2000) Ibid.

[19]Dickenson, D. (2002) Commodification of human tissue: implications for feminist and development ethics, *Developing World Bioethics,* **2(1)** 55–63; 59.

[20]Radin, M. J. (1996) *Contested commodities: The trouble with trade in sex, children, body parts and other things,* Harvard University Press, Cambridge, MA.

[21]Anderson, E. (1993) *Value in Ethics and Economics,* Harvard University Press, Cambridge, MA.

[22]Anderson (1993) *op. cit.,* p. 175.

[23]Anderson (1993) *op. cit.*, p. 170.

[24]Holland (2001) *op. cit.*, pp. 278–280.

[25]Holland (2001) *op. cit.*, p. 279.

[26]Radin, M. J. (2001) Response: persistent perplexities, *Kennedy Inst. Ethics J.* **11(3)** 305–315, p. 310.

[27]Dickenson, D. (2001) Property and women's alienation from their own reproductive labour, *Bioethics* **15(3)** 205–217.

[28]Dickenson, (2002), *op. cit.*, p. 56.

[29] Marx, K. (1867, 1954) *Capital*, Engels, F, ed., Moore, S. and Aveling, E., trans., Progress Press, Moscow, pp. 167, 170.

[30]Delphy, C. (1984) *Close to Home: A Materialist Analysis of Women's Oppression*, Leonard, D., trans., Hutchinson, London; cited in Dickenson (2001).

[31]Dephy (1984) *op. cit.*, cited in Dickenson (2001), p. 209.

[32]Dickenson (2002) *op. cit.*, p. 61

[33]Dickenson (2001) p. 213.

[34]Dickenson (2001) pp. 213–214.

[35]Further criticisms can be made of Dickenson's reading of Marx on alienation via the *Grundrisse* and her interpretation of Marx's distinction between what is natural and what is social, but they are not central to the argument here.

[36]Cohen. G. (1995) *Self-ownership, Freedom and Equality*, Cambridge University Press, New York, p. 144ff.

[37]Marx (1867/1954) *op. cit.*, p. 43.

[38]As Dickenson points out (2002), p. 57 at law tissues and organs that have been removed from the body are deemed to be no-one's property. Cf. Moore v Regents of the University of California (1990) 51 Cal.3d 120, which reaffirmed this legal doctrine preventing a man from obtaining any portion of the financial benefits gained from a cell line derived (without his consent) from his T-cells.

[39]Marx (1867/1954) *op. cit.*, pp. 164-65.

[40]Marx (1867/1954) *op. cit.*, pp. 165.

[41]Marx (1867/1954) *op. cit.*, pp. 166.

[42]*See*, for example, Mackenzie, C. (1992) Abortion and embodiment, *Australas. J. Philos.*, **70**, 136–155, in which she queries the adequacy of approaches to the ethics of abortion which narrowly construe pregnancy as an event in the lives of women over which women lack control.

Index